Routledge Revivals

The Geometry of Vector Fields

This volume, first published in 2000, presents a classical approach to the foundations and development of the geometry of vector fields, describing vector fields in three-dimensional Euclidean space, triply-orthogonal systems and applications in mechanics. Topics covered include Pfaffian forms, systems in n-dimensional space, and foliations and their Godbillion-Vey invariant. There is much interest in the study of geometrical objects in n-dimensional Euclidean space and this volume provides a useful and comprehensive presentation.

The Geometry of Vector Fields

Yu. Aminov

ROUTLEDGE REVIVALS
Routledge
Taylor & Francis Group

First published in 2000
by Gordon and Breach Science Publishers

This edition first published in 2013 by Routledge
2 Park Square, Milton Park, Abingdon, Oxon, OX14 4RN

Simultaneously published in the USA and Canada
by Routledge
711 Third Avenue, New York, NY 10017

Routledge is an imprint of the Taylor & Francis Group, an informa business

Publisher's Note
The publisher has gone to great lengths to ensure the quality of this reprint but points out that some imperfections in the original copies may be apparent.

Disclaimer
The publisher has made every effort to trace copyright holders and welcomes correspondence from those they have been unable to contact.

A Library of Congress record exists under LC control number: 2002421350

ISBN 13: 978-0-415-70685-8 (hbk)
ISBN 13: 978-1-315-88723-4 (ebk)

The Geometry of Vector Fields

Yu. Aminov
Institute for Low Temperature Physics and Engineering
Kharkov, Ukraine

GORDON AND BREACH SCIENCE PUBLISHERS
Australia • Canada • France • Germany • India
Japan • Luxembourg • Malaysia • The Netherlands
Russia • Singapore • Switzerland

Amsteldijk 166
1st Floor
1079 LH Amsterdam
The Netherlands

British Library Cataloguing in Publication Data

A catalogue record for this book is available from the British Library.

ISBN: 90-5699-201-5

Dedicated to

my parents

Contents

Preface

For a regular family of surfaces given in a domain of three-dimensional Euclidean space one can discover geometrical properties of this family from the very fact that the surfaces totally fill the domain and lie in space in a good way, i.e. without mutual intersections. These properties cannot be reduced to the geometry of the surfaces themselves but are connected with their spatial configuration. We can study these properties by considering the vector field of unit normals on the surfaces. Many surface metric invariants can be found with the help of normal vector fields. Therefore, the geometry of a vector field or, in other terminology, the geometry of the distribution of planes orthogonal to the vector field, i.e. non-holonomic geometry, includes the theory of families of surfaces. The set of all vector fields is wider than the set of normal vector fields and our study of arbitrary vector fields in contrast to the fields of normals of surfaces is necessary, for various reasons. Many results concerning the family of surfaces stay valid, as well, for an arbitrary vector field and this particular case makes these results clearer. It is possible to introduce the concepts of Gaussian and mean curvatures, the two analogues of geodesic lines and lines of curvature, and the analogues of asymptotic lines for the case of a vector field. Also, a vector field has its own geometrical invariant which shows how much a given vector field differs from the field of normals of a family of surfaces. This quantity we call the *non-holonomicity value* of a vector field.

Besides the study of the distribution of planes orthogonal to a given field, it is natural to study the behavior of the streamlines of a vector field from the metric viewpoint. In this context the geometry of a vector field is closely related to the theory of ordinary differential equations.

It is useful to provide some background information on the development of the geometry of vector fields and the mathematicians associated with this study. The foundations of the geometry of vector fields were proposed by A. Voss at the end of the nineteenth century and were followed by the publication of papers by Lie,

Carathéodory, Liliental and Darboux. A. Voss defined the notions of lines of curvature, asymptotic lines, and he also considered some special classes of vector fields and mechanical systems with non-holonomic constraints. Liliental studied the analogue of geodesic lines — the shortest lines orthogonal to the field. Rogers gave the interpretation of the non-holonomicity value in terms of mean geodesic torsion. Later on, the study of the geometry of vector fields was developed in papers by D. Sintsov, Schouten, van Kampen, Rogers, J. Blank, V. Vagner, G. Vransceanu and others. D. Sintsov produced the generalization of the Beltrami–Enneper formula, stated the relations between the curvatures of the first and second kind, considered the lines of curvature, and introduced the notion of the indicatrix of geodesic torsion, and his works have been published as a monograph [16]. J. Blank considered vector fields from the projective point of view.

The papers by Vagner on non-holonomic geometry were awarded the Lobachevski prize. Vagner introduced the notion of parallel transport with respect to a vector field, and generalized the Gauss–Bonnet formula to the case of a strictly non-holonomic vector field. In this monograph we present these results in simple language, avoiding the difficult methods in the Vagner paper. Another generalization of this formula, based on the concept of spherical image, which works for the case of an arbitrary vector field in n-dimensional Euclidean space, was found by the author. In the preparation of this book, in addition to the papers of other mathematicians, the author's results on 'divergent' properties of symmetric functions of principal curvatures, on the behavior of streamlines, and on complex non-holonomicity have been included. We also present the author's result on mutual restrictions of the fundamental field invariants and the size of the domain of definition of the field. These results belong to 'geometry in the large'. One can find their origins in papers by N. Efimov on the geometry of surfaces of negative curvature.

An important area of applications of non-holonomic geometry is the dynamics of mechanical systems with non-holonomic constraints. Usually, these constraints arise in the description of the rolling of a rigid body over the surface of another body, taking friction into account. A. Voss, S. Chaplygin and other scientists generalized the Lagrange system of equations to the case of non-holonomic constraints. S. Chaplygin studied the rolling of a ball-like body over the plane in detail. From the geometrical viewpoint, mechanical systems with non-holonomic constraints have been considered by V. Vagner and G. Vransceanu. Currently, systems with non-holonomic constraints are studied intensively in mechanics ([51], [96], [97]).

The second field of applications is the geometry of velocity fields of fluid flow. From the literature on this subject we highlight the papers by S. Bushgens, which present necessary and sufficient conditions for the vector field to be the field of velocity directions of the stationary flow of an ideal incompressible fluid.

Vector fields occur in a natural manner also in general relativity. Recently it has become clear that vector fields of constant length are used in the description of liquid crystals and ferromagnets. It is interesting to note that the Hopf invariant, involved in the description of vector fields defined in space with the point at infinity included, is also used by physicists, although they use the term 'topological charge' [76].

The contents of this book are divided naturally into two chapters. Chapter 1 is devoted to vector fields in three–dimensional Euclidean space, to triply-orthogonal systems and to applications in mechanics. In Chapter 2 vector fields, Pfaffian forms and systems in n-dimensional space, foliations and their Godbillon–Vey invariant are considered. We also present the connection stated by Whitehead between the integrated non-holonomicity value and the Hopf invariant.

The study of geometrical objects in n-dimensional Euclidean space is an exciting problem. The notions introduced here have descriptive interpretations and the theory can be demonstrated easily by concrete examples of vector fields. The methods include no complicated or, more precisely, awkward mathematical tools. Nevertheless, there are some problems where complexity is essential. Note that some important applied investigations lead to vector fields in E^n.

I would like to convey my appreciation to V. M. Bykov for his useful and encouraging positive remarks.

I would also like to express my special thanks to Professor A. Yampol'ski for the translation and to M. Goncharenko for her technical assistance.

1 Vector Fields in Three-Dimensional Euclidean Space

1.1 The Non-Holonomicity Value of a Vector Field

Let us consider a family of surfaces defined in some domain G of three-dimensional Euclidean space E^3. The family of surfaces can be determined in different ways. For instance, we can regard the family surface as a level surface $\Phi(x_1, x_2, x_3) = \text{const}$ of some function $\Phi(x_1, x_2, x_3)$ which depends on three arguments x_1, x_2, x_3. We suppose that only one surface of the family passes through each point M in the domain G, i.e. the family under consideration is regular. Then at each point M in G there is a well-defined unit vector $\mathbf{n}$ — the normal vector of the surface which passes through that point. Therefore, the vector function $\mathbf{n} = \mathbf{n}(M)$ is defined in the domain G. Let x_1, x_2, x_3 be Cartesian coordinates in E^3 and $\mathbf{e}_1, \mathbf{e}_2, \mathbf{e}_3$ be unit basis vectors. Then we can regard the vector field $\mathbf{n}$ as the vector function of the parameters x_1, x_2, x_3. This means that each component ξ_i of $\mathbf{n}$ is a function of three Cartesian coordinates x_1, x_2, x_3 of a point M in G:

$$\xi_i = \xi_i(x_1, x_2, x_3), \quad i = 1, 2, 3.$$

The unit vector field $\mathbf{n}$, being orthogonal to the family of surfaces, is not arbitrary. Construct another vector field $\operatorname{curl}\mathbf{n}$ by means of the vector field $\mathbf{n}$ as follows:

$$\operatorname{curl}\mathbf{n} = \mathbf{e}_1(\xi_{3x_2} - \xi_{2x_3}) + \mathbf{e}_2(\xi_{1x_3} - \xi_{3x_1}) + \mathbf{e}_3(\xi_{2x_1} - \xi_{1x_2}).$$

We may represent it in symbolic form as

$$\operatorname{curl}\mathbf{n} = \begin{vmatrix} \mathbf{e}_1 & \mathbf{e}_2 & \mathbf{e}_3 \\ \frac{\partial}{\partial x_1} & \frac{\partial}{\partial x_2} & \frac{\partial}{\partial x_3} \\ \xi_1 & \xi_2 & \xi_3 \end{vmatrix}.$$

We shall need the following formulas in the sequel. If $\boldsymbol{v}$, $\mathbf{a}$ and $\mathbf{b}$ are some vector fields and λ is a function of x_i then

$$\begin{aligned} \operatorname{curl} \lambda \boldsymbol{v} &= \lambda \operatorname{curl} \boldsymbol{v} + [\operatorname{grad} \lambda, \boldsymbol{v}], \\ \operatorname{curl}[\mathbf{ab}] &= \mathbf{a} \operatorname{div} \mathbf{b} - \mathbf{b} \operatorname{div} \mathbf{a} + \nabla_{\mathbf{b}} \mathbf{a} - \nabla_{\mathbf{a}} \mathbf{b}, \end{aligned} \tag{1}$$

where $\nabla_{\mathbf{b}}\mathbf{a}$ means the derivative of $\mathbf{a}$ via $\mathbf{b}$, i.e. the vector of components $\left\{\sum_\alpha \frac{\partial a_i}{\partial x_\alpha} b_\alpha\right\}$.

The following theorem holds.

Theorem (Jacobi) *If* $(\mathbf{n}, \operatorname{curl} \mathbf{n}) = 0$ *at each point then there is a family of surfaces orthogonal to the vector field* $\mathbf{n}$ *and vice versa.*

Let $(\mathbf{n}, \operatorname{curl} \mathbf{n}) = 0$ in a domain G. Show that there is a family of surfaces $f(x_1, x_2, x_3) = \text{const}$ orthogonal to the vector field $\mathbf{n}$.

Consider at some point $M_0 \in G$ of coordinates (x_1^0, x_2^0, x_3^0) the set of directions $dr = \{dx_1, dx_2, dx_3\}$ each of which is orthogonal to $\mathbf{n}$, i.e. those directions which satisfy the equation

$$\xi_1 dx_1 + \xi_2 dx_2 + \xi_3 dx_3 = 0. \tag{2}$$

We may assume that the system of coordinates in E^3 has been chosen in such a way that at M_0 and, as a consequence, in some its neighborhood $\xi_3 \neq 0$. We can rewrite equation (2) as follows:

$$dx_3 = -\frac{\xi_1}{\xi_3} dx_1 - \frac{\xi_2}{\xi_3} dx_2. \tag{3}$$

We try to find a function $x_3 = z(x_1, x_2)$ which satisfies equation (3) and such that $z(x_1^0, x_2^0) = x_3^0$. In this case dx_3 is the differential of that function and the coefficients of dx_1 and dx_2 in (3) are the partial derivatives of the function $z(x_1, x_2)$, i.e. the equation (3) is equivalent to the following system:

$$\frac{\partial z}{\partial x_1} = -\frac{\xi_1}{\xi_3}, \quad \frac{\partial z}{\partial x_2} = -\frac{\xi_2}{\xi_3}. \tag{4}$$

The right-hand sides of these equations depend on x_1, x_2 and z. System (4) is solvable if the compatibility condition

$$\left(\frac{\xi_1}{\xi_3}\right)_{x_2} - \left(\frac{\xi_2}{\xi_3}\right)_{x_1} = 0 \tag{5}$$

is satisfied, where one ought to find the derivatives taking into account (4). We have

$$\begin{aligned} \left(\frac{\xi_1}{\xi_3}\right)_{x_2} &= \left[\left(\xi_{1x_2} + \xi_{1x_3}\frac{\partial z}{\partial x_2}\right)\xi_3 - \left(\xi_{3x_2} + \xi_{3x_3}\frac{\partial z}{\partial x_2}\right)\xi_1\right]\xi_3^{-2} \\ &= (\xi_{1x_2}\xi_3^2 - \xi_{1x_3}\xi_2\xi_3 + \xi_{3x_3}\xi_1\xi_2 - \xi_{3x_2}\xi_3\xi_1)\xi_3^{-3}. \end{aligned}$$

Here we replaced z_{x_2} with $-\xi_2/\xi_3$ by (4). Interchanging the roles of indices 1 and 2, we get

$$\left(\frac{\xi_2}{\xi_3}\right)_{x_1} = (\xi_{2x_1}\xi_3^2 - \xi_{2x_3}\xi_1\xi_3 - \xi_{3x_1}\xi_3\xi_2 + \xi_{3x_3}\xi_1\xi_2)\xi_3^{-3}.$$

The difference of expressions above produces the compatibility condition

$$\begin{aligned}\left(\frac{\xi_1}{\xi_3}\right)_{x_2} - \left(\frac{\xi_2}{\xi_3}\right)_{x_1} &= [\xi_1(\xi_{2x_3} - \xi_{3x_2}) + \xi_2(\xi_{3x_1} - \xi_{1x_3}) + \xi_3(\xi_{1x_2} - \xi_{2x_1})]\xi_3^{-2} \\ &= -(\mathbf{n}, \operatorname{curl}\mathbf{n})\xi_3^{-2} = 0. \qquad (6)\end{aligned}$$

Since the latter equation is satisfied by hypothesis, there is a function $z = z(x_1, x_2, x_3^0)$, where x_3^0 is a fixed parameter, which satisfies the system (4) and such that $z(x_1^0, x_2^0, x_3^0) = x_3^0$. This function defines the surface of the position vector $\mathbf{r} = \{x_1, x_2, z(x_1, x_2, x_3^0)\}$, where x_3^0 is fixed. Due to the condition $z(x_1^0, x_2^0, x_3^0) = x_3^0$ this surface passes through the point M_0. Varying x_3^0 we obtain a family of surfaces satisfying equation (2) at each point. That equation means that the tangent plane of the surface is orthogonal to **n**.

Conversely, if the family of surfaces orthogonal to **n** exists then equation (2) is satisfied and for the surface which passes through the point M_0 it is possible to consider one of parameters, say x_3, as a function of x_1 and x_2. Then system (4) will be satisfied and the compatibility condition (5) will be fulfilled. Due to (6) this means that $(\mathbf{n}, \operatorname{curl}\mathbf{n}) = 0$.

So, *the vector field* **n** *which is orthogonal to the family of surfaces is special and is called holonomic*. The field **n** holonomicity condition can be represented in terms of non-collinear vector fields **a** and **b** which are orthogonal to **n**. Let $[\mathbf{a}, \mathbf{b}] = \lambda\mathbf{n}$, where $\lambda = |[\mathbf{a}, \mathbf{b}]| \neq 0$. Applying formula (1), we can write

$$\begin{aligned}(\mathbf{n}, \operatorname{curl}\mathbf{n}) &= \lambda^{-1}([\mathbf{a}, \mathbf{b}], \operatorname{curl}\lambda^{-1}[\mathbf{a}, \mathbf{b}]) = \lambda^{-2}([\mathbf{a}, \mathbf{b}], \operatorname{curl}[\mathbf{a}, \mathbf{b}]) \\ &= \lambda^{-2}([\mathbf{a}, \mathbf{b}], \mathbf{a}\operatorname{div}\mathbf{b} - \mathbf{b}\operatorname{div}\mathbf{a} + \nabla_{\mathbf{b}}\mathbf{a} - \nabla_{\mathbf{a}}\mathbf{b}) \\ &= \lambda^{-1}(\mathbf{n}, \nabla_{\mathbf{b}}\mathbf{a} - \nabla_{\mathbf{a}}\mathbf{b}).\end{aligned}$$

From this it follows that $(\mathbf{n}, \operatorname{curl}\mathbf{n}) = 0$ if and only if the vector $\nabla_{\mathbf{b}}\mathbf{a} - \nabla_{\mathbf{a}}\mathbf{b}$ is orthogonal to **n**. The vector $\nabla_{\mathbf{b}}\mathbf{a} - \nabla_{\mathbf{a}}\mathbf{b}$ is called the *Poisson bracket* and is denoted by $\nabla(\mathbf{a}, \mathbf{b})$.

It is natural to expand the class of vector fields under consideration, avoiding the requirement $(\mathbf{n}, \operatorname{curl}\mathbf{n}) = 0$. We shall call the value $\rho = (\mathbf{n}, \operatorname{curl}\mathbf{n})$ *the value of non-holonomicity* of a vector field **n**. It is independent of the choice of direction of **n**. Since **n** is a unit vector field, the value ρ has some geometrical sense. We give two geometrical interpretations of the non-holonomicity value ρ.

1. *The Vagner interpretation*. Let M_0 be some point in a domain G where the vector field **n** is defined. Draw a plane through M_0 which is orthogonal to $\mathbf{n}(M_0)$ and a circle L in this plane such that L passes through M_0. Set that positive orientation in L to be defined by $\mathbf{n}(M_0)$. The field **n** is defined in some neighborhood of M_0

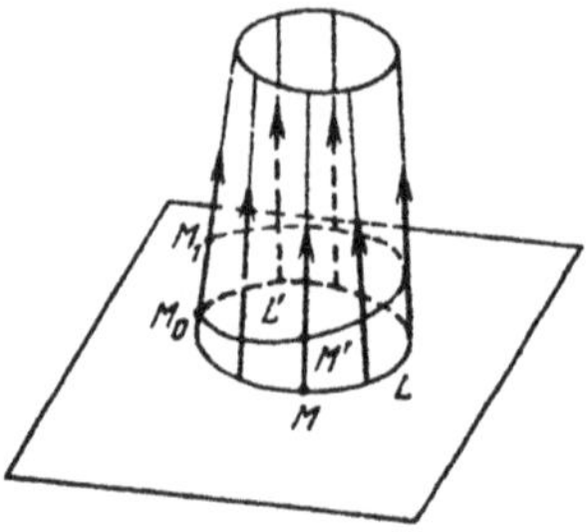

FIGURE 1

including the points of L. Draw a straight line through each point M in L in the direction of $\mathbf{n}(M)$. We obtain some ruled surface. Starting from M_0, draw a curve in this surface which is orthogonal to the elements (see Fig. 1).

Consider one coil of that curve which is one-to-one projectable with the elements onto the positively oriented L. Denote it by L'. The starting point of L' is at M_0. The curve L' meets the straight line element through M_0 at another point M_1. Denote by l the length of M_1M_0 and by σ the area of disk bounded by L. Then the value of non-holonomicity of a vector field $\mathbf{n}$ at M_0 is equal to the limit of ratio $-l/\sigma$ when L tends to M_0:

$$(\mathbf{n}, \operatorname{curl} \mathbf{n})|_{M_0} = -\lim_{L \to M_0} \frac{l}{\sigma}. \tag{7}$$

We find the limit of l/σ under assumption that the radius of circle L tends to zero. Let M be an arbitrary point in L and s be the arc length of L between M_0 and M_1 in the positive direction (see Fig. 1). Let M' be a point in L' projecting onto M in L. Denote the distance between M and M' by $u(s)$. Let S be the length of the circle L. Show that there is a positive constant C such that for all circles which pass through M_0 and lie in some neighborhood of M_0 the following inequality holds:

$$|u(s)| \leq S^2 C. \tag{8}$$

To prove this, represent a position vector of L' in the form

$$\mathbf{r}(s) = \mathbf{r}_1(s) + u(s)\,\mathbf{n}(s),$$

where $\mathbf{r}_1(s)$ is a position vector of L and $\mathbf{n}(s)$ is the field vector at the points of L. At the points of L' the equality

$$(\mathbf{r}', \mathbf{n}(s)) = 0$$

must be satisfied. We have

$$\mathbf{r}' = \mathbf{r}_1' + u'\mathbf{n} + u\,\mathbf{n}'.$$

Note that $\mathbf{r}_1'$ is a unit tangent vector of L. Multiplying by $\mathbf{n}$, we get

$$u' = -(\mathbf{r}_1', \mathbf{n}) = -\cos\varphi,$$

where φ is the angle between $\mathbf{n}$ and the tangent to L. So,

$$u(s) = -\int_0^s \cos\varphi\, ds.$$

Denote by α the angle between $\mathbf{n}(M)$ and the plane where L is located. The angle φ is greater than α and, as a consequence, $|\cos\varphi| \leq \cos\alpha$. The angle α is determined by the vector field $\mathbf{n}$. We can regard α as a function of coordinates x, y in the plane which contains L. Moreover, $\alpha(0,0) = \pi/2$. Then we are able to write an expression

$$\cos\alpha = (\cos\alpha)_x \Delta x + (\cos\alpha)_y \Delta y + \mathbf{o}(\Delta r_1),$$

where $\mathbf{o}(\Delta r_1)$ is the infinitesimal of a higher order than $\Delta r_1 = \sqrt{\Delta x^2 + \Delta y^2}$. From this we get

$$|\cos\varphi| \leq |\cos\alpha| \leq C_1 \Delta r_1 \leq C_2 S.$$

So,

$$|u(s)| \leq \int_0^s |\cos\alpha|\, ds \leq \frac{C_2}{2} S^2 = CS^2,$$

where C, C_1 and C_2 are constants.

Let L'' be a closed curve formed with L' and the intercept M_1M_0. Consider the circulation of $\mathbf{n}$ along the closed curves L and L''. They enclose a ruled surface F. Denote its area by σ'. By Stokes' theorem we have

$$\int_L (\mathbf{n}, d\mathbf{r}) - \int_{L''} (\mathbf{n}, d\mathbf{r}) = \iint_F (\operatorname{curl} \mathbf{n}, \nu)\, d\sigma,$$

where ν is the normal to F, $d\sigma$ is the area element. The following inequality holds:

$$\left| \iint_F (\operatorname{curl} \mathbf{n}, \nu)\, d\sigma \right| \leq \max_G |\operatorname{curl} \mathbf{n}| \sigma'.$$

As the vector field $\mathbf{n}$ is regular, the derivatives of $\mathbf{n}$ and, as a consequence, $|\operatorname{curl}\mathbf{n}|$ are bounded from above in G with some number. The area of ruled surface bounded by L and L'' can be estimated from above using (8):

$$\sigma' = \int_L |u(s)| ds \leq S^3 C_3.$$

Therefore, the difference of circulations of **n** along L and L'' is of the order S^3:

$$\left|\int_L (\mathbf{n}, d\mathbf{r}) - \int_{L''} (\mathbf{n}, d\mathbf{r})\right| \le C_4 S^3,$$

where $C_4 = \text{const}$. Let Φ be a disk bounded by L. The area of Φ is $\sigma = S^2/4\pi$. That is why

$$\lim_{S\to 0}\left|\int_L (\mathbf{n}, d\mathbf{r}) - \int_{L''} (\mathbf{n}, d\mathbf{r})\right|\sigma^{-1} \le 4\pi C_4 \lim_{S\to 0}\frac{S^3}{S^2} = 0.$$

Hence

$$\lim_{S\to 0}\frac{1}{\sigma}\int_L (\mathbf{n}, d\mathbf{r}) = \lim_{S\to 0}\frac{1}{\sigma}\int_{L''} (\mathbf{n}, d\mathbf{r}). \tag{9}$$

Now find the circulation of **n** along L''. Write an expansion of the vector function **n** along each straight line element of the ruled surface

$$\mathbf{n}(M') = \mathbf{n}(M) + \frac{\partial n}{\partial \tau}(M)u(s) + \mathbf{o}(u(s)), \tag{10}$$

where $\partial/\partial\tau$ is the derivative of the vector field **n** at M in the direction of $\mathbf{n}(M)$, $\mathbf{o}(u(s))$ is the infinitesimal of a higher order than the infinitesimal $u(s)$. We can write

$$\int_{L''} (\mathbf{n}, d\mathbf{r}) = \int_{L'} (\mathbf{n}(M'), d\mathbf{r}) + \int_{M_1M_0} (\mathbf{n}, d\mathbf{r}).$$

In the first integral on the right we may replace $\mathbf{n}(M')$ by (10). As $\mathbf{n}(M)$ is orthogonal to $d\mathbf{r}$ along L' this integral has an upper bound

$$C\int_l |u(s)|\,ds \le CS^3.$$

That is why its quotient to σ tends to zero, when $S \to 0$. Analogously,

$$\int_{M_1M_0} (\mathbf{n}, d\mathbf{r}) \approx -l, \tag{11}$$

moreover, this integral differs from $-l$ by an infinitesimal of the order S^3.

The circulation of **n** along the circle L can also be evaluated by means of a double integral over the disk Φ. We have

$$\iint_\Phi (\operatorname{curl}\mathbf{n}, \mathbf{n}(M_0))\,d\sigma = \int_L (\mathbf{n}\,dr) \approx \sigma(\mathbf{n}, \operatorname{curl}\mathbf{n})|_{M_0}.$$

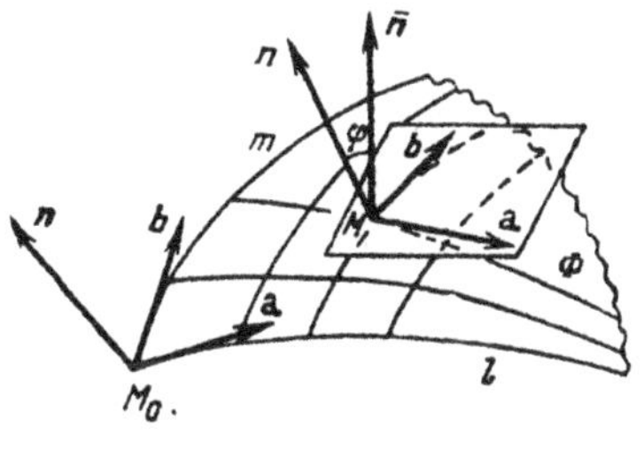

FIGURE 2

Divide both sides by σ and pass to the limit, when $S \to 0$. Taking into account (9) and (11) we obtain (7).

2. *The second interpretation of the non-holonomicity value* geometrically answers the question of why it is impossible to construct the surface orthogonal to $\mathbf{n}$ in the non-holonomic case. Construct the vector fields $\mathbf{a}$ and $\mathbf{b}$ which are orthogonal to each other and $\mathbf{n}$ in some neighborhood of a point M_0. To do this we can take, for instance, a constant vector field ξ which is not collinear to $\mathbf{n}(M_0)$ and set $\mathbf{a} = [\mathbf{n}, \xi]/|[\mathbf{n}, \xi]|$. Then $\mathbf{b} = [\mathbf{n}, \mathbf{a}]$. Draw a curve l through M_0 which has $\mathbf{a}$ as a tangent vector field. Then draw a curve m through M_0 which has $\mathbf{b}$ as a tangent vector field. Starting from the points of m, emit the lines tangentially to the vectors of the field $\mathbf{a}$. They form some surface Φ (see Fig. 2). Denote by $\bar{\mathbf{n}}(M)$ a unit normal of that surface at its arbitrary point M. The $\bar{\mathbf{n}}(M)$ is orthogonal to $\mathbf{a}(M)$, hence it is located in the plane spanned by $\mathbf{n}(M)$ and $\mathbf{b}(M)$. Denote by φ the angle between $\bar{\mathbf{n}}(M)$ and $\mathbf{n}(M)$. Then at M_0 and the other points of m

$$\frac{d\varphi}{ds} = -(\mathbf{n}, \operatorname{curl} \mathbf{n}), \tag{12}$$

where d/ds is the derivative with respect to the arc length parameter of the curve tangential to the field $\mathbf{a}$.

To prove (12) consider the decomposition

$$\bar{\mathbf{n}}(M) = \cos\varphi \mathbf{n}(M) + \sin\varphi \mathbf{b}(M).$$

Prolongate the vector field $\bar{\mathbf{n}}(M)$ into some neighborhood of the surface Φ. Since the surface Φ which is orthogonal to $\bar{\mathbf{n}}$ exists by construction, then $(\bar{\mathbf{n}}, \operatorname{curl} \bar{\mathbf{n}}) = 0$.

We can write

$$\operatorname{curl} \bar{\mathbf{n}} = \cos\varphi \operatorname{curl} \mathbf{n} + \sin\varphi \operatorname{curl} \mathbf{b} + [\mathbf{n}, \operatorname{grad}\varphi] \sin\varphi - [\mathbf{b}, \operatorname{grad}\varphi] \cos\varphi.$$

Therefore

$$\begin{aligned}(\bar{\mathbf{n}}, \operatorname{curl} \bar{\mathbf{n}}) = {} & \cos^2\varphi(\mathbf{n}, \operatorname{curl} \mathbf{n}) + \sin^2\varphi(\mathbf{b}, \operatorname{curl} \mathbf{b}) \\ & - (\mathbf{n}, \mathbf{b}, \operatorname{grad}\varphi) + \cos\varphi \sin\varphi \{(\mathbf{b}, \operatorname{curl} \mathbf{n}) + (\mathbf{n}, \operatorname{curl} \mathbf{b})\} = 0.\end{aligned}$$

Since $[\mathbf{n}, \mathbf{b}] = -\mathbf{a}$, then from the latter equation we find

$$\frac{d\varphi}{ds} = -(\mathbf{n}, \operatorname{curl} \mathbf{n}) + \sin^2\varphi\{(\mathbf{n}, \operatorname{curl} \mathbf{n}) - (\mathbf{b}, \operatorname{curl} \mathbf{b})\}$$
$$- \cos\varphi \sin\varphi\{(\mathbf{b}, \operatorname{curl} \mathbf{n}) + (\mathbf{n}, \operatorname{curl} \mathbf{b})\}. \tag{13}$$

At each point of m the normal $\bar{\mathbf{n}}(M)$ of the surface Φ coincides with $\mathbf{n}$. Therefore at the points of that curve $\varphi = 0$. From (13) it follows that at the points of m we have (12).

Formula (12) shows that for a non-holonomic vector field $\mathbf{n}$ the surface which is orthogonal to $\mathbf{n}$ does not exist. Indeed, let us suppose that at M_0 and, as a consequence, in some part of its neighborhood $(\mathbf{n}, \operatorname{curl} \mathbf{n}) \neq 0$. A surface which is orthogonal to $\mathbf{n}$ and passes through M_0 must contain the curves l and m because they are orthogonal to $\mathbf{n}$. Hence, it must contain curves tangential to $\mathbf{a}$ emitted from the points of m because the field $\mathbf{a}$ is orthogonal to $\mathbf{n}$, i.e. that surface must coincide with Φ and its normal field $\bar{\mathbf{n}}$ must coincide with $\mathbf{n}$. But from (12) it follows that $\frac{d\varphi}{ds} \neq 0$ in some neighborhood of M_0. Therefore, in some neighborhood of M_0 we have $\varphi \neq 0$, i.e. the normal field of Φ does not coincide with the given field $\mathbf{n}$.

At the end of this section we give an example of a vector field with a constant non-zero non-holonomicity value. Set

$$\xi_1 = r\cos\varphi, \quad \xi_2 = r\sin\varphi, \quad \xi_3 = \sqrt{1 - r^2},$$

where r is positive constant, $\varphi = \varphi(x_3)$. We have

$$\operatorname{curl} \mathbf{n} = \{-r\cos\varphi\varphi_{x_3}, -r\sin\varphi\varphi_{x_3}, 0\}.$$

Hence,

$$(\mathbf{n}, \operatorname{curl} \mathbf{n}) = -r^2\varphi_{x_3}.$$

If $\varphi = ax_3 + b$ with constant a and b then the value of non-holonomicity $(\mathbf{n}, \operatorname{curl} \mathbf{n}) = -r^2 a = \text{const}$.

1.2 Normal Curvature of a Vector Field and Principal Normal Curvatures of the First Kind

The surface behavior about a point M_0 on it can be described with the help of the surface normal curvature k_n which can be found for any surface tangent direction τ. Geometrically, k_n is interpreted as a curvature of surface normal section at M_0 in the direction of τ. Analytically, k_n can be found as a ratio of the surface second fundamental form $II = (d^2\mathbf{r}, \mathbf{n})$ to the first one $I = d\mathbf{r}^2$:

$$k_n = \frac{(d^2\mathbf{r}, \mathbf{n})}{(d\mathbf{r}, d\mathbf{r})},$$

where $\mathbf{r}$ is a position vector of the surface, $\mathbf{n}$ is a unit normal at M_0. Since $\mathbf{n}$ is a vector of unit length, it is possible to rewrite the expression above as

$$k_n = -\frac{(d\mathbf{r}, d\mathbf{n})}{(d\mathbf{r}, d\mathbf{r})}.$$

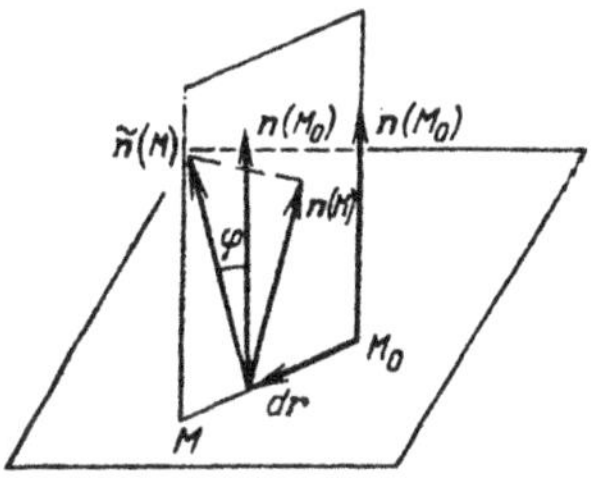

FIGURE 3

Suppose now that an arbitrary regular unit vector field **n** is defined in a domain G. We shall introduce a curvature of the field (i.e. a function which makes **n** different from a constant vector field) at M_0. In the holonomic case the curvature will coincide with the normal curvature of the surface which is orthogonal to the field.

Draw a plane through M_0 which is orthogonal to $\mathbf{n}(M_0)$. Let $M_0M = d\mathbf{r}$ be a shift in this plane. Draw another plane through $\mathbf{n}(M_0)$ and $d\mathbf{r}$. Let $\tilde{\mathbf{n}}(M)$ be a projection of $\mathbf{n}(M)$ into the latter plane (see Fig. 3).

Denote by φ the angle between $\mathbf{n}(M_0)$ and $\tilde{\mathbf{n}}(M)$. The *normal curvature* $k_n(d\mathbf{r})$ *(or simply* k_n*) of a vector field* **n** *in the direction* $d\mathbf{r}$ is the limit of the ratio $-\varphi/|d\mathbf{r}|$ when $M \to M_0$, i.e.

$$k_n(d\mathbf{r}) = \lim_{M \to M_0} \frac{-\varphi}{|d\mathbf{r}|}. \tag{1}$$

If **n** is a constant vector field then $\varphi = 0$ at any point M and, as a consequence, $k_n = 0$.

Prove that in the holonomic case we come to the definition of the surface normal curvature. As a sine of φ is equal to the cosine of an angle between $\tilde{\mathbf{n}}(M)$ and $d\mathbf{r}$, we see that

$$\sin\varphi = \frac{(\tilde{\mathbf{n}}(M), d\mathbf{r})}{|\tilde{\mathbf{n}}||d\mathbf{r}|}. \tag{2}$$

As the infinitesimals φ and $\sin\varphi$ are of the same order, it is possible to replace φ with $\sin\varphi$ in (1). As $\tilde{\mathbf{n}}(M)$ is the projection of $\mathbf{n}(M)$ into the plane of vectors $\mathbf{n}(M_0)$ and $d\mathbf{r}$,

$$(\tilde{\mathbf{n}}(M), d\mathbf{r}) = (\mathbf{n}(M), d\mathbf{r}).$$

In the neighborhood of M_0 it is possible to produce the following expression:

$$\mathbf{n}(M) = \mathbf{n}(M_0) + d\mathbf{n} + \mathbf{o}(d\mathbf{r}),$$

where $|\mathbf{o}(d\mathbf{r})|$ is the infinitesimal of a higher order than $|d\mathbf{r}|$. Therefore,

$$(\tilde{\mathbf{n}}(M), d\mathbf{r}) = (d\mathbf{n}, d\mathbf{r}) + (\mathbf{o}(d\mathbf{r}), d\mathbf{r}). \tag{3}$$

Note that $|\tilde{\mathbf{n}}| \to 1$ when $M \to M_0$. Using (2) and (3) we write

$$k_n(d\mathbf{r}) = \lim_{M \to M_0} \frac{-(\tilde{\mathbf{n}}(M), d\mathbf{r})}{|\tilde{\mathbf{n}}||d\mathbf{r}|^2} = -\frac{(d\mathbf{n}, d\mathbf{r})}{d\mathbf{r}^2}.$$

This expression coincides with the expression of the surface normal curvature.

The normal curvature of a vector field depends on the direction $d\mathbf{r}$. If we rotate $d\mathbf{r}$ in the plane which is orthogonal to $\mathbf{n}(M_0)$ then it varies, in general, and reaches its extremal values. These extremal normal curvatures are called the *principal curvatures of the first kind* . The corresponding directions are called the *principal directions of the first kind.* In analogy with the theory of surfaces, the curves tangent to these principal directions are called the *lines of curvature of the first kind.* Find the equations on the principal directions and lines of curvature. To do this write the expression of k_n in terms of components of $\mathbf{n}$:

$$\begin{aligned} k_n = & -\left[\xi_{1x_1} dx_1^2 + (\xi_{1x_2} + \xi_{2x_1})\, dx_1\, dx_2 + \xi_{2x_2} dx_2^2 \right. \\ & + (\xi_{1x_3} + \xi_{3x_1})\, dx_1\, dx_3 + (\xi_{2x_3} + \xi_{3x_2})\, dx_2\, dx_3 \\ & \left. + \xi_{3x_3} dx_3^2\right] (dx_1^2 + dx_2^2 + dx_3^2)^{-1}. \end{aligned}$$

Choose the coordinate axes in E^3 in such a way that $\xi_1 = \xi_2 = 0$ at M_0. Since $\xi_3 = \sqrt{1 - \xi_1^2 - \xi_2^2}$, $\xi_{3x_i} = 0$ at M_0 for $i = 1, 2, 3$. The value of k_n does not depend on the length of $d\mathbf{r}$. So, set $|d\mathbf{r}| = 1$. We shall find the extremal values of k_n provided that $d\mathbf{r}$ is in the plane orthogonal $\mathbf{n}(M_0)$, i.e. $d\mathbf{r} = (dx_1, dx_2, 0)$. In these supposed conditions k_n takes the form

$$k_n = -\left[\xi_{1x_1} dx_1^2 + (\xi_{1x_2} + \xi_{2x_1})\, dx_1\, dx_2 + \xi_{2x_2} dx_2^2\right]. \tag{4}$$

As $dx_1^2 + dx_2^2 = 1$, we use the Lagrange multiplier method to find the extremal values. Taking dx_1 and dx_2 as independent variables, we want to find the extremal values of the following function:

$$\xi_{1x_1} dx_1^2 + (\xi_{1x_2} + \xi_{2x_1})\, dx_1\, dx_2 + \xi_{2x_2} dx_2^2 + \lambda(dx_1^2 + dx_2^2).$$

This leads us to the following system of equations with respect to the principal directions:

$$\begin{aligned} (\xi_{1x_1} + \lambda) dx_1 + \frac{\xi_{1x_2} + \xi_{2x_1}}{2} dx_2 = 0, \\ \frac{\xi_{1x_2} + \xi_{2x_1}}{2} dx_1 + (\xi_{2x_2} + \lambda) dx_2 = 0. \end{aligned} \tag{5}$$

The characteristic equation for λ has the form

$$\lambda^2 + \lambda(\xi_{1x_1} + \xi_{2x_2}) + \xi_{1x_1}\xi_{2x_2} - \left(\frac{\xi_{1x_2} + \xi_{2x_1}}{2}\right)^2 = 0.$$

The matrix of system (5) is symmetric, so that we have two real roots λ_1 and λ_2. Just as in the theory of surfaces, it is possible to prove that these roots are the principal curvatures of the field and the principal directions of the first kind are mutually orthogonal. The *mean curvature of the field* is a half-sum of principal curvatures:

$$H = \frac{1}{2}(\lambda_1 + \lambda_2) = -\frac{1}{2}(\xi_{1x_1} + \xi_{2x_2}). \tag{6}$$

The *total curvature of the first kind* is a product of principal curvatures:

$$K_1 = \lambda_1\lambda_2 = \xi_{1x_1}\xi_{2x_2} - \left(\frac{\xi_{1x_2} + \xi_{2x_1}}{2}\right)^2. \tag{7}$$

A point M_0 is called elliptic, parabolic or hyperbolic if at this point $K_1 > 0$, $K_1 = 0$, $K_1 < 0$ respectively.

Let us find the equations of lines of curvature of the first kind. Rewrite system (5) in the form

$$\begin{aligned} &d\xi_1 + \frac{\xi_{2x_1} - \xi_{1x_2}}{2}\,dx_2 + \lambda\,dx_1 = 0,\\ &d\xi_2 + \frac{\xi_{1x_2} - \xi_{2x_1}}{2}\,dx_2 + \lambda\,dx_2 = 0. \end{aligned} \tag{8}$$

Note that $\xi_{2x_1} - \xi_{1x_2}$ is a third component of curl **n**. If we set $(\operatorname{curl}\mathbf{n})_i$ to be the i-th component of **n** then we can write (8) as

$$2(d\xi_1 + \lambda\,dx_1) + (\operatorname{curl}\mathbf{n})_3\,dx_2 - (\operatorname{curl}\mathbf{n})_2\,dx_3 = 0,$$
$$2(d\xi_2 + \lambda\,dx_2) + (\operatorname{curl}\mathbf{n})_1\,dx_3 - (\operatorname{curl}\mathbf{n})_3\,dx_1 = 0,$$

or as

$$2(d\xi_i + \lambda\,dx_i) + [d\mathbf{r}, \operatorname{curl}\mathbf{n}]_i = 0, \quad i = 1, 2, \tag{9}$$

where $[\ ]_i$ means the i-th component of the vector product. System (9) shows that a projection of

$$2(d\mathbf{n} + \lambda\,d\mathbf{r}) + [d\mathbf{r}, \operatorname{curl}\mathbf{n}] \tag{10}$$

into the plane orthogonal to **n** is zero, i.e. the vector (10) is collinear with **n**. Therefore, the triple of vectors $2\,d\mathbf{n} + [d\mathbf{r}, \operatorname{curl}\mathbf{n}]$, $d\mathbf{r}$ and **n** is linearly dependent and, as a consequence, their mixed product is zero:

$$2(d\mathbf{n}, d\mathbf{r}, \mathbf{n}) + ([d\mathbf{r}, \operatorname{curl}\mathbf{n}], d\mathbf{r}, \mathbf{n}) = 0.$$

Using the formula $[[\mathbf{a}, \mathbf{b}], \mathbf{c}] = (\mathbf{a}, \mathbf{c})\mathbf{b} - \mathbf{a}(\mathbf{b}, \mathbf{c})$ and the orthogonality property of $d\mathbf{r}$ and **n** we obtain the system of the equation which determines the lines of curvature of the first kind:

$$\begin{aligned} 2(d\mathbf{n}, d\mathbf{r}, \mathbf{n}) + (\mathbf{n}, \operatorname{curl}\mathbf{n})d\mathbf{r}^2 &= 0,\\ (\mathbf{n}, d\mathbf{r}) &= 0. \end{aligned} \tag{11}$$

As $d\mathbf{n} = \sum_i \mathbf{n}_{x_i}\,dx_i$, we can rewrite system (11) as

$$\sum_{i,j=1}^{3} A_{ij}\,dx_i\,dx_j = 0,$$

$$\xi_1\,dx_1 + \xi_2\,dx_2 + \xi_3\,dx_3 = 0,$$

where A_{ij} are some coefficients depending on **n** and its derivatives. Take one of the coordinates, say x_3, as a parameter t of the curve. From the second equation we find

$$\frac{dx_2}{dt} = -\frac{\xi_1}{\xi_2}\frac{dx_1}{dt} - \xi_3.$$

Substitute this into the first equation. Solving the latter quadratic equation with respect to dx_1/dt, we find

$$\frac{dx_k}{dt} = F_k(\xi_i, \xi_{ix_j}), \quad k = 1, 2,$$

$$\frac{dx_3}{dt} = 1.$$

The right-hand sides of the equations above depend on x_1, x_2, x_3. Integrating that system we can find the required lines of curvature.

The directions of principal normal curvatures of the first kind are mutually orthogonal. Indeed, returning to system (5), note that the required directions are the eigenvectors of the symmetric matrix

$$\left\| \begin{matrix} \xi_{1x_1} & \frac{\xi_{1x_2}+\xi_{2x_1}}{2} \\ \frac{\xi_{1x_2}+\xi_{2x_1}}{2} & \xi_{2x_2} \end{matrix} \right\|$$

From this the analog of the Euler theorem follows. Take the principal directions as the coordinate axes. Set $dx_1/ds = \cos\alpha$, $dx_2/ds = \sin\alpha$. From the assumption $dx_3 = 0$ and (4) it follows that

$$k_n = -(\xi_{1x_1}\cos^2\alpha + (\xi_{1x_2} + \xi_{2x_1})\cos\alpha\,\sin\alpha + \xi_{2x_2}\sin^2\alpha).$$

From system (5), we find

$$(\xi_{1x_2} + \xi_{2x_1})(dx_2^2 - dx_1^2) + 2(\xi_{1x_2} - \xi_{2x_1})\,dx_1\,dx_2 = 0.$$

Due to the choice of coordinate axes the solutions of the latter equations are $(dx_1, 0)$ and $(0, dx_2)$, hence $\xi_{1x_2} + \xi_{2x_1} = 0$. Setting $\alpha = 0,\ \pi/2$ we find that $-\xi_{1x_1}$ and $-\xi_{2x_2}$ are the principal curvatures k_1 and k_2 of the first kind.

Therefore, the Euler formula for the case of a vector field has the same form as in the theory of surfaces:

$$k_n = k_1\cos^2\alpha + k_2\sin^2\alpha.$$

1.3 The Streamline of Vector Field

The streamline of vector field **n** is a line which is tangential to **n** at each of its points (see Fig. 4). Let (ξ_1, ξ_2, ξ_3) be the components of the vector field **n** with respect to the

FIGURE 4

Cartesian system of coordinates (x_1, x_2, x_3). Then the streamline must satisfy the system of differential equations

$$\frac{dx_i}{ds} = \xi_i(x_1, x_2, x_3), \quad i = 1, 2, 3.$$

We introduce the Hamilton formula for the vector of curvature of the streamline of a given field **n**. The required vector is directed along the streamline principal normal and its length is equal to the streamline curvature. We denote it by **k**. The *Hamilton formula* has the following form:

$$\mathbf{k} = -[\mathbf{n}, \operatorname{curl} \mathbf{n}].$$

To prove this it is sufficient to show, for instance, that the first components of **k** and $-[\mathbf{n}, \operatorname{curl} \mathbf{n}]$ coincide.

If we denote the differentiation with respect to the **n** streamline arc length parameter by d/ds then

$$\mathbf{k} = \frac{d\mathbf{n}}{ds} = \mathbf{n}_{x_i}\xi_i.$$

Therefore, the first component of **k** is equal to $\xi_{1x_i}\xi_i$. Consider the first component of $-[\mathbf{n}, \operatorname{curl} \mathbf{n}]$. It is equal to

$$\begin{aligned}\xi_2(\xi_{1x_2} - \xi_{2x_1}) - \xi_3(\xi_{3x_1} - \xi_{1x_3}) &= \xi_{1x_2}\xi_2 + \xi_{1x_3}\xi_3 - \xi_{2x_1}\xi_2 - \xi_{3x_1}\xi_3 \\ &= \xi_{1x_i}\xi_i - \frac{1}{2}\frac{\partial}{\partial x_1}(\xi_1^2 + \xi_2^2 + \xi_3^2) \\ &= \xi_{1x_i}\xi_i,\end{aligned}$$

which completes the proof. From this it follows that for the case of a field of directions with congruent straight lines we have $-[\mathbf{n}, \operatorname{curl} \mathbf{n}] = 0$, i.e. curl **n** is collinear with **n**.

Let us take a family of level surfaces $U(x_1, x_2, x_3) = \text{const}$. Let **n** be the field of normals. Find the **n** streamline curvature vector **k** in this case, i.e. the orthogonal trajectories of surfaces family. Let $\mathbf{n} = \{\frac{dx_i}{ds}\}$ be the unit tangent vector of the streamline. Set $H = |\text{grad}\, U|^{-1}$. Then

$$\frac{dx_i}{ds} = \frac{\partial U}{\partial x_i} H, \quad i = 1, 2, 3.$$

The first component of $\mathbf{k}$ is equal to

$$\frac{d^2 x_1}{ds^2} = \frac{\partial}{\partial x_i}\left(H \frac{\partial U}{\partial x_1}\right)\frac{dx_i}{ds} = \frac{\partial H}{\partial x_i}\frac{\partial U}{\partial x_1}\frac{\partial U}{\partial x_i} H + \frac{\partial^2 U}{\partial x_1 \partial x_i}\frac{\partial U}{\partial x_i} H^2.$$

Note that $H_{x_1} = -\frac{\partial^2 U}{\partial x_1 \partial x_i}\frac{\partial U}{\partial x_i} H^3$. Therefore, the following is true:

$$\frac{d^2 x_1}{ds^2} = -\frac{H_{x_1}}{H} + \frac{(\text{grad}\, H, \mathbf{n})}{H}\frac{dx_1}{ds}.$$

Analogous calculations with respect to the second and third components of $\mathbf{k}$ show that

$$\mathbf{k} = -\text{grad} \ln H + \mathbf{n}(\text{grad} \ln H, \mathbf{n}).$$

The expression above was stated by Darboux.

1.4 The Straightest and the Shortest Lines

The geodesic lines in a surface possess two properties of straight lines in a plane: they are the straightest lines (that is, their geodesic curvature is equal to zero) and shortest local lines simultaneously (that is, the geodesic line has the smallest length among all curves joining two sufficiently close points P_1 and P_2).

In the case of a non-holonomic vector field we can determine the straightest and the shortest lines among the curves which are orthogonal to the field.

Let the field $\mathbf{n}$ be orthogonal to the curve Γ. Let $\mathbf{r} = \mathbf{r}(s)$ be a position vector of Γ, where s is the arc length parameter. The geodesic curvature $1/\rho_g$ of Γ with respect to the field $\mathbf{n}$ is the length of the projection of the curvature vector of Γ, i.e. of $\mathbf{r}_{ss}$, into the plane orthogonal to $\mathbf{n}$. We set it positive if the triple $(\mathbf{r}_s, \mathbf{n}, \mathbf{r}_{ss})$ is positively oriented and negative in the opposite case, i.e. we set

$$\frac{1}{\rho_g} = -(\mathbf{n}, \mathbf{r}_s, \mathbf{r}_{ss}).$$

Set $\mathbf{r}_s = \mathbf{e}$. Suppose that the unit vector field $\mathbf{e}$ is defined in a three-dimensional domain. Then $\text{curl}\, \mathbf{e}$ is also defined and $\mathbf{r}_{ss} = -[\mathbf{e}, \text{curl}\, \mathbf{e}]$ by Hamilton's formula. The geodesic curvature with respect to the field $\mathbf{n}$ of lines tangent to $\mathbf{e}$ takes the form

$$\frac{1}{\rho_g} = -(\mathbf{n}, \text{curl}\, \mathbf{e}).$$

In the case of the straightest line the geodesic curvature with respect to **n** is equal to zero. Therefore, that curve satisfies the equation

$$(\mathbf{n}, \mathbf{r}_s, \mathbf{r}_{ss}) = 0. \tag{1}$$

Equation (1) means that

$$\mathbf{r}_{ss} = \mu \mathbf{r}_s + \lambda \mathbf{n}.$$

As s is the arc length parameter, $(\mathbf{r}_s, \mathbf{r}_{ss}) = 0$, i.e. $\mu = 0$. Hence, the straightest line curvature vector is directed along **n**:

$$\mathbf{r}_{ss} = \lambda \mathbf{n}.$$

The coefficient λ is equal to the normal curvature of the field in the direction $\mathbf{r}_s$, i.e. $\lambda = (\mathbf{r}_{ss}, \mathbf{n}) = -(\mathbf{r}_s, \mathbf{n})$. Therefore the vector equation of the straightest line has the form

$$\mathbf{r}_{ss} = -(\mathbf{r}_s, \mathbf{n}_s)\mathbf{n} = -(\mathbf{r}_s, \mathbf{n}_{x_i})\,\mathbf{n}\frac{dx_i}{ds}. \tag{2}$$

This is the differential equation of second order with respect to the vector function $\mathbf{r}(s)$ parametrized with some parameter s. If the curve is orthogonal to **n** at the initial point P_0 then it stays orthogonal to **n** for all s. Indeed, the function $(\mathbf{r}_s, \mathbf{n})$ satisfies the following equation:

$$(\mathbf{r}_s, \mathbf{n})_s = (\mathbf{r}_{ss}, \mathbf{n}) + (\mathbf{r}_s, \mathbf{n}_s) = 0,$$

i.e. $(\mathbf{r}_s, \mathbf{n}) = \text{const}$. As $(\mathbf{r}_s, \mathbf{n}) = 0$ at P_0 then $(\mathbf{r}_s, \mathbf{n}) \equiv 0$. If $\mathbf{r}(s)$ is a solution of (2) then take the initial condition for $\mathbf{r}_s$ such that $|\mathbf{r}_s| = 1$. Then $|\mathbf{r}_s| \equiv 1$ and, as a consequence, s is the arc length parameter. It is sufficient to show that $\mathbf{r}_s^2 = \text{const}$. We have

$$(\mathbf{r}_s^2)_s = 2(\mathbf{r}_s, \mathbf{r}_{ss}) = 2(\mathbf{r}_s, \mathbf{n})\lambda = 0.$$

The solution of (2) is determined by the initial conditions $\mathbf{r}(0)$, $\mathbf{r}_s(0)$.

Thus, we can draw a unique straightest line in the given direction $\mathbf{r}_s(0)$ which is orthogonal to the field through any point of the domain G. We call the straightest line the geodesic.

Turn now to the consideration of shortest lines.

Let the points P_0 and P_1 be joined with the curves orthogonal to the field **n**. *We call those curves admissible.* Among all admissible curves we shall find a curve with the smallest length among all nearly admissible ones.

We consider a position vector **r** of the shortest line L as a function of the arc length parameter s which varies from 0 to l. We vary the curve L in such a way that it will stay admissible and the curve coordinate variations $\delta x_1, \delta x_2, \delta x_3$ turn into zero at

P_0 and P_1. Take the parameter s as a parameter in the varied curve. Let dx_i/ds be a tangent vector of the varied curve. Then the following equation must be satisfied:

$$\xi_i \frac{dx_i}{ds} = 0. \tag{3}$$

The shortest line L gives the extremal value of the integral

$$\int_0^l ds = \int_L \sqrt{dx_1^2 + dx_2^2 + dx_3^2}$$

among all curves which satisfy (3). Applying the Lagrange method, we find the conditional extremum or, equivalently, the curve which gives the ordinary extremum of the following integral:

$$I = \int_0^l \left\{ 1 + \varphi(s)\xi_i \frac{dx_i}{ds} \right\} ds,$$

where $\varphi(s)$ is an unknown function, provided that $\xi_i \frac{dx_i}{ds} = 0$ at the points of the required curve L.

Consider the variation of $\xi_i \frac{dx_i}{ds}$ in an arbitrary direction $\delta\mathbf{r} = (\delta x_1, \delta x_2, \delta x_3)$. We shall consider the admissible curves as lying on the surface $\mathbf{r} = \mathbf{r}(s, \alpha)$ provided that for any fixed α we get the admissible curve. Then

$$\frac{\partial}{\partial\alpha}\xi_i \frac{\partial x_i}{\partial s} = \frac{\partial \xi_i}{\partial \alpha}\frac{\partial x_i}{\partial s} + \xi_i \frac{\partial^2 x_i}{\partial s \partial \alpha}. \tag{4}$$

Set $-\mathrm{curl}\,\mathbf{n} = \{l_i\}$:

$$l_1 = \xi_{2x_3} - \xi_{3x_2}, \qquad l_2 = \xi_{3x_1} - \xi_{1x_3}, \qquad l_3 = \xi_{1x_2} - \xi_{2x_1},$$

or using briefer notation

$$\frac{\partial \xi_i}{\partial x_j} - \frac{\partial \xi_j}{\partial x_i} = l_\alpha \varepsilon^{ij\alpha}$$

where $\varepsilon^{ij\alpha}$ is the Kronecker symbol. Then we can rewrite (4) as

$$\begin{aligned}
\frac{\partial}{\partial\alpha}\xi_i \frac{\partial x_i}{\partial s} &= \frac{\partial \xi_i}{\partial x_j}\frac{\delta x_j}{\delta\alpha}\frac{\partial x_i}{\partial s} + \xi_i \frac{\partial^2 x_i}{\partial s \partial \alpha} \\
&= \left(l_\alpha \varepsilon^{ij\alpha} + \frac{\partial \xi_j}{\partial x_i} \right) \frac{\delta x_j}{\delta \alpha}\frac{\partial x_i}{\partial s} + \xi_i \frac{\partial^2 x_i}{\partial s \partial \alpha} \\
&= \xi_i \frac{\partial^2 x_i}{\partial s \partial \alpha} + \varepsilon^{ij\alpha} l_\alpha \frac{\delta x_j}{\delta\alpha}\frac{\partial x_i}{\partial s} + \frac{d\xi_j}{ds}\frac{\delta x_j}{\delta\alpha} \\
&= \xi_i \frac{\partial^2 x_i}{\partial s \partial \alpha} - \left(\mathrm{curl}\,\mathbf{n}, \frac{d\mathbf{r}}{ds}, \frac{\delta\mathbf{r}}{\delta\alpha} \right) + \left(\frac{d\mathbf{n}}{ds}, \frac{\delta\mathbf{r}}{\delta\alpha} \right).
\end{aligned} \tag{5}$$

Using (3) and (5), we get

$$\frac{\delta I}{\delta\alpha} = \int_L \frac{\frac{dx_i}{ds}\frac{\partial}{\partial\alpha}\frac{dx_i}{ds}\,ds}{\sqrt{\sum\left(\frac{dx_i}{ds}\right)^2}} + \int_L \varphi(s)\left\{\xi_i\frac{\partial^2 x_i}{\partial s\partial\alpha} - \left(\operatorname{curl}\mathbf{n}, \frac{d\mathbf{r}}{ds}, \frac{\delta\mathbf{r}}{\delta\alpha}\right) + \left(\frac{d\mathbf{n}}{ds}, \frac{\delta\mathbf{r}}{\delta\alpha}\right)\right\}ds$$

$$= \int_L \left\{-\xi_i\frac{\partial x_i}{\partial\alpha}\frac{d\varphi(s)}{ds} - \left(\operatorname{curl}\mathbf{n}, \frac{d\mathbf{r}}{ds}, \frac{\delta\mathbf{r}}{\delta\alpha}\right)\varphi - \frac{\delta x_i}{\delta\alpha}\frac{d^2x_i}{ds^2}\right\}ds.$$

The condition $\delta I/\delta\alpha = 0$ and the expression above imply

$$\frac{d^2\mathbf{r}}{ds^2} + \mathbf{n}\frac{d\varphi}{ds} + \varphi(s)\left[\operatorname{curl}\mathbf{n}, \frac{d\mathbf{r}}{ds}\right] = 0. \tag{6}$$

Our next aim is to exclude the function $\varphi(s)$. Take the mutually orthogonal unit vectors $\mathbf{n}$, $\frac{d\mathbf{r}}{ds}$, $\left[\mathbf{n}, \frac{d\mathbf{r}}{ds}\right]$ as the basis along L. Decompose $\operatorname{curl}\mathbf{n}$ and $\frac{d^2\mathbf{r}}{ds}$ with respect to that basis. We have

$$\operatorname{curl}\mathbf{n} = (\mathbf{n}, \operatorname{curl}\mathbf{n})\mathbf{n} + a\left[\mathbf{n}, \frac{d\mathbf{r}}{ds}\right] + b\frac{d\mathbf{r}}{ds}. \tag{7}$$

We can represent the curvature vector of the shortest line L as

$$\frac{d^2\mathbf{r}}{ds^2} = \frac{1}{R_1}\left[\mathbf{n}, \frac{d\mathbf{r}}{ds}\right] + \frac{1}{R_2}\mathbf{n}. \tag{8}$$

Substituting (7) and (8) into (6), we get

$$\frac{1}{R_2}\left[\mathbf{n}, \frac{d\mathbf{r}}{ds}\right] + \frac{1}{R_2}\mathbf{n} + \mathbf{n}\frac{d\varphi}{ds} + \varphi(s)(\mathbf{n}, \operatorname{curl}\mathbf{n})\left[\mathbf{n}, \frac{d\mathbf{r}}{ds}\right] + \varphi(s)a\left[\left[\mathbf{n}, \frac{d\mathbf{r}}{ds}\right], \frac{d\mathbf{r}}{ds}\right] = 0.$$

Setting the coefficients of $\mathbf{n}$ and $\left[\mathbf{n}, \frac{d\mathbf{r}}{ds}\right]$ equal to zero, we obtain

$$\frac{1}{R_1} + \varphi(s)(\mathbf{n}, \operatorname{curl}\mathbf{n}) = 0, \tag{9}$$

$$\frac{1}{R_2} + \frac{d\varphi}{ds} - \varphi a = 0. \tag{10}$$

In the case of a holonomic vector field we have $(\mathbf{n}, \operatorname{curl}\mathbf{n}) = 0$ and the equation of a geodesic line obtains the form

$$\frac{1}{R_1} = -\left(\frac{d^2\mathbf{r}}{ds^2}, \frac{d\mathbf{r}}{ds}, \mathbf{n}\right) = 0,$$

i.e. in this case the geodesic curvature along the shortest line is zero.

Suppose that $(\mathbf{n}, \operatorname{curl}\mathbf{n}) \neq 0$ along some part of L. Then the function $\varphi(s)$ can be excluded from these two equations. Resolving (9) with respect to $\varphi(s)$ and substituting into (10), we obtain

$$\frac{1}{R_2} - \frac{d}{ds}\frac{1}{R_1(\mathbf{n}, \operatorname{curl}\mathbf{n})} - \frac{a}{R_1(\mathbf{n}, \operatorname{curl}\mathbf{n})} = 0.$$

From (5) and (6) we obtain the expression for $1/R_2$ and a. Namely

$$\frac{1}{R_2} = \left(\frac{d^2\mathbf{r}}{ds^2}, \mathbf{n}\right) = -\left(\frac{d\mathbf{r}}{ds}, \frac{d\mathbf{n}}{ds}\right),$$

i.e. $1/R_2$ is the field normal curvature in the direction of L. Denote it by k_n. Also,

$$a = \left(\operatorname{curl}\mathbf{n}, \mathbf{n}, \frac{d\mathbf{r}}{ds}\right) = \left(\mathbf{k}, \frac{d\mathbf{r}}{ds}\right),$$

i.e. $-a$ with the opposite sign is equal to the projection of the field streamline curvature vector onto the tangent of L. So, the equation of the shortest line has the form

$$\frac{d}{ds}\frac{\left(\frac{d^2\mathbf{r}}{ds^2}, \frac{d\mathbf{r}}{ds}, \mathbf{n}\right)}{(\mathbf{n}, \operatorname{curl}\mathbf{n})} - \left(\frac{d\mathbf{r}}{ds}, \frac{d\mathbf{n}}{ds}\right) + \left(\mathbf{k}, \frac{d\mathbf{r}}{ds}\right)\frac{\left(\frac{d^2\mathbf{r}}{ds^2}, \frac{d\mathbf{r}}{ds}, \mathbf{n}\right)}{(\mathbf{n}, \operatorname{curl}\mathbf{n})} = 0.$$

Substituting $\varphi = -\frac{1}{R_1(\mathbf{n}, \operatorname{curl}\mathbf{n})}$ into (6), we obtain the vector form for the equation of the shortest line:

$$\frac{d^2\mathbf{r}}{ds^2} + \mathbf{n}\frac{d}{ds}\frac{\left(\frac{d^2\mathbf{r}}{ds^2}, \frac{d\mathbf{r}}{ds}, \mathbf{n}\right)}{(\mathbf{n}, \operatorname{curl}\mathbf{n})} + \frac{\left(\frac{d^2\mathbf{r}}{ds^2}, \frac{d\mathbf{r}}{ds}, \mathbf{n}\right)}{(\mathbf{n}, \operatorname{curl}\mathbf{n})}\left[\operatorname{curl}\mathbf{n}, \frac{d\mathbf{r}}{ds}\right] = 0. \tag{12}$$

This is the third-order differential equation with respect to the position vector $\mathbf{r}(s)$.

Write the projections of the right and left sides of the latter equation with respect to movable axes adjoint to the field $\mathbf{n}$. Let $\mathbf{a}$ and $\mathbf{b}$ be the unit mutually orthogonal vector fields such that $[\mathbf{a}, \mathbf{b}] = \mathbf{n}$. Set

$$\frac{d\mathbf{r}}{ds} = \cos\alpha\mathbf{a} + \sin\alpha\mathbf{b}.$$

Therefore

$$\frac{d^2\mathbf{r}}{ds^2} = (-\sin\alpha\mathbf{a} + \cos\alpha\mathbf{b})\frac{d\alpha}{ds} + \cos\alpha\frac{d\mathbf{a}}{ds} + \sin\alpha\frac{d\mathbf{b}}{ds}.$$

We can represent the derivatives of $\mathbf{a}$ and $\mathbf{b}$ with respect to the arc length parameter in terms of derivatives with respect to the Cartesian coordinates x_i:

$$\frac{d\mathbf{a}}{ds} = \mathbf{a}_{x_i}\frac{dx_i}{ds}, \qquad \frac{d\mathbf{b}}{ds} = \mathbf{b}_{x_i}\frac{dx_i}{ds}.$$

As $\mathbf{a}$ and $\mathbf{b}$ are of unit length, then $\mathbf{a}_{x_i}$ is orthogonal to $\mathbf{a}$ and $\mathbf{b}_{x_i}$ is orthogonal to $\mathbf{b}$. Consider the decompositions

$$\mathbf{a}_{x_i} = L_i\mathbf{b} + T_i\mathbf{n},$$
$$\mathbf{b}_{x_i} = N_i\mathbf{a} + M_i\mathbf{n}.$$

As **a** and **b** are mutually orthogonal, then $L_i = -N_i$. Find $\left(\frac{d^2\mathbf{r}}{ds^2}, \frac{d\mathbf{r}}{ds}, \mathbf{n}\right)$. We have

$$\left(\frac{d^2\mathbf{r}}{ds^2}, \frac{d\mathbf{r}}{ds}, \mathbf{n}\right) = -\frac{d\alpha}{ds} + \cos^2\alpha\left(\frac{d\mathbf{a}}{ds}, \mathbf{a}, \mathbf{n}\right) + \sin^2\alpha\left(\frac{d\mathbf{b}}{ds}, \mathbf{b}, \mathbf{n}\right)$$

$$= -\left(\frac{d\alpha}{ds} + L_i(\cos\alpha\mathbf{a}_i + \sin\alpha\mathbf{b}_i)\right), \tag{13}$$

where $\mathbf{a}_i$ and $\mathbf{b}_i$ are Cartesian components of vector fields **a** and **b** respectively. Multiply (12) by **n** scalarly. We obtain

$$\left(\frac{d^2\mathbf{r}}{ds^2}, \mathbf{n}\right) + \frac{d}{ds}\frac{\left(\frac{d^2\mathbf{r}}{ds^2}, \frac{d\mathbf{r}}{ds}, \mathbf{n}\right)}{(\mathbf{n}, \operatorname{curl}\mathbf{n})} + \frac{\left(\frac{d^2\mathbf{r}}{ds^2}, \frac{d\mathbf{r}}{ds}, \mathbf{n}\right)}{(\mathbf{n}, \operatorname{curl}\mathbf{n})}\left(\mathbf{n}, \operatorname{curl}\mathbf{n}, \frac{d\mathbf{r}}{ds}\right) = 0. \tag{14}$$

We have

$$\left(\frac{d^2\mathbf{r}}{ds^2}, \mathbf{n}\right) = \cos\alpha\left(\frac{d\mathbf{a}}{ds}, \mathbf{n}\right) + \sin\alpha\left(\frac{d\mathbf{b}}{ds}, \mathbf{n}\right)$$

$$= \cos^2\alpha A + \cos\alpha\sin\alpha B + \sin^2\alpha C, \tag{15}$$

$$\left(\mathbf{n}, \operatorname{curl}\mathbf{n}, \frac{d\mathbf{r}}{ds}\right) = (\mathbf{n}, \operatorname{curl}\mathbf{n}, \cos\alpha\mathbf{a} + \sin\alpha\mathbf{b})$$

$$= \cos\alpha P + \sin\alpha Q, \tag{16}$$

where A, B, C, P and Q are some functions of x_i. Using (13), (15) and (16), we can represent equation (14) as

$$-\frac{d}{ds}\frac{\left(\frac{d\alpha}{ds} + L_i(\cos\alpha\mathbf{a}_i + \sin\alpha\mathbf{b}_i)\right)}{(\mathbf{n}, \operatorname{curl}\mathbf{n})} + \cos^2\alpha A + \cos\alpha\sin\alpha B + \sin^2\alpha C$$

$$-\frac{\cos\alpha P + \sin\alpha Q}{(\mathbf{n}, \operatorname{curl}\mathbf{n})}\left(\frac{d\alpha}{ds} + L_i(\cos\alpha\mathbf{a}_i + \sin\alpha\mathbf{b}_i)\right) = 0. \tag{17}$$

Consider the projection of equation (12) onto **a**. We have

$$\left(\frac{d^2\mathbf{r}}{ds^2}, \mathbf{a}\right) = -\sin\alpha\frac{d\alpha}{ds} + \sin\alpha\left(\frac{d\mathbf{b}}{ds}, \mathbf{a}\right)$$

$$= -\sin\alpha\left(\frac{d\alpha}{ds} + L_i\frac{dx_i}{ds}\right).$$

Comparing with (13), we get

$$\left(\frac{d^2\mathbf{r}}{ds^2}, \mathbf{a}\right) = \sin\alpha\left(\frac{d^2\mathbf{r}}{ds^2}, \frac{d\mathbf{r}}{ds}, \mathbf{n}\right). \tag{18}$$

Next we have

$$\left(\mathbf{a},\ \operatorname{curl}\mathbf{n},\ \frac{d\mathbf{r}}{ds}\right) = (\mathbf{a}\ \operatorname{curl}\mathbf{n},\ \cos\alpha\mathbf{a} + \sin\alpha\mathbf{b})$$
$$= -\sin\alpha(\mathbf{n},\ \operatorname{curl}\mathbf{n}). \qquad (19)$$

Taking into account (18) and (19), we conclude that the equation obtained as a projection of (12) onto **a** is satisfied identically. The same is true for the projection of (12) onto **b**. Thus, only equation (17) is left which can be rewritten in general form

$$\frac{d}{ds}\left(\frac{d\alpha/ds}{(\mathbf{n},\operatorname{curl}\mathbf{n})}\right) = F\left(x_1, x_2, x_3, \alpha, \frac{d\alpha}{ds}\right). \qquad (20)$$

To this equation we associate the equation

$$\frac{d\mathbf{r}}{ds} = \cos\alpha\mathbf{a} + \sin\alpha\mathbf{b}. \qquad (21)$$

The equations (20), (21) form a system of ordinary differential equations with respect to the functions $x_i(s)$, $i = 1, 2, 3$, $\alpha = \alpha(s)$. Let P_0 be a point of the coordinates $x_i(0)$ such that $(\mathbf{n}, \operatorname{curl}\mathbf{n}) \neq 0$ at this point. Suppose that this point corresponds to the value $s = 0$. Give the initial values to the unknown functions at $s = 0$: $x_i(0)$, $i = 1, 2, 3$ $\alpha(0)$, $\frac{d\alpha}{ds}(0)$. The choice of $x_i(0)$ means the choice of the starting point of the shortest line, the choice of $\alpha(0)$ gives the direction of the shortest line at P_0. Thus, the following theorem holds.

Theorem *If* $(\mathbf{n}, \operatorname{curl}\mathbf{n}) \neq 0$ *at some point* P_0 *then a bundle of shortest lines with respect to the given direction passes through this point.*

Give an example of a vector field **n** with a simple structure but non-constant with respect to which the equation (17) is integrable in quadratures. Suppose the field **n** is parallel to the plane of x_1, x_2, i.e. it has the following components:

$$\xi_1 = \cos\varphi, \qquad \xi_2 = \sin\varphi, \qquad \xi_3 = 0.$$

For the sake of simplicity, we suppose that $\varphi = cx_3 + d$, where c, d are constants. Then we may set

$$\mathbf{a} = e_3, \qquad \mathbf{b} = -\sin\varphi\mathbf{e}_1 + \cos\varphi\mathbf{e}_2.$$

We see that

$$\operatorname{curl}\mathbf{n} = \{-c\cos\varphi, -c\sin\varphi, 0\} = -c\,\mathbf{n}.$$

Therefore, $(\mathbf{n}, \operatorname{curl}\mathbf{n}) = -c$, $[\mathbf{n}, \operatorname{curl}\mathbf{n}] = 0$. Since **a** is a constant vector, $L_i = T_i = 0$,

$$\left(\frac{d^2\mathbf{r}}{ds^2}, \frac{d\mathbf{r}}{ds}, \mathbf{n}\right) = -\frac{d\alpha}{ds}.$$

Let us find $\left(\frac{d^2\mathbf{r}}{ds^2}, \mathbf{n}\right)$ for this case. Differentiating **b**, we find

$$\frac{d\mathbf{b}}{ds} = -\mathbf{n}\frac{d\varphi}{ds} = -\mathbf{n}c\frac{dx_3}{ds}.$$

From the equation

$$\frac{d\mathbf{r}}{ds} = \cos\alpha\mathbf{a} + \sin\alpha\mathbf{b}$$

we find $dx_3/ds = \cos\alpha$. Hence, $d\mathbf{b}/ds = -\mathbf{n}c\cos\alpha$. We rewrite equation (15) as

$$\left(\frac{d^2\mathbf{r}}{ds^2}, \mathbf{n}\right) = -\sin\alpha\cos\alpha c.$$

Therefore, equation(14) of the shortest line has the form

$$\frac{d^2\alpha}{ds^2} = -c^2\sin\alpha\cos\alpha.$$

Suppose that $d\alpha/ds \neq 0$. Multiplying by $2d\alpha/ds$ and integrating, we find

$$\left(\frac{d\alpha}{ds}\right)^2 = c^2\cos^2\alpha + c_1,$$

where c_1 is a constant. We find the function $\alpha(s)$ from the following equation:

$$s = \int \frac{d\alpha}{\sqrt{c^2\cos^2\alpha + c_1}}.$$

To find a position vector of the shortest line we use the representation of $d\mathbf{r}/ds$ via $\mathbf{a}$ and $\mathbf{b}$, which in coordinate form is

$$\frac{dx_1}{ds} = -\sin\alpha\sin\varphi, \quad \frac{dx_2}{ds} = \sin\alpha\cos\varphi, \quad \frac{dx_3}{ds} = \cos\alpha, \tag{22}$$

where the dependence of φ from s and α is given by

$$d\varphi = -c\cos\alpha\, ds = -\frac{c\cos\alpha\, d\alpha}{\sqrt{c\cos^2\alpha + c_1}}.$$

Therefore $\sin(\varphi + c_2) = -c\sin\alpha/\sqrt{c^2 - c_1}$. One of the solutions of (21) is evident. This is a family of straight lines parallel to the x_3 axis.

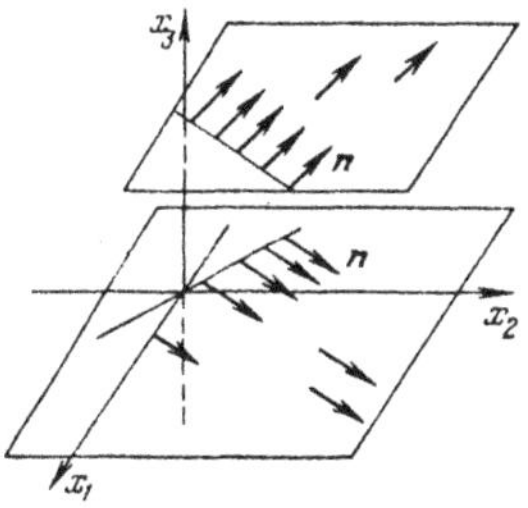

FIGURE 5

As in each plane of $x_3 = \text{const}$ the field $\mathbf{n}$ is constant, the other solution is a family of straight lines in these planes which are orthogonal to $\mathbf{n}$ (see Fig. 5).

1.5 The Total Curvature of the Second Kind

In this section we introduce another notion of the Gaussian curvature of a vector field analogous to the Gaussian curvature of a surface. Consider at some point M_0 a small piece of plane orthogonal to $\mathbf{n}(M_0)$. Take some neighborhood of M_0 in this plane and map it into the unit sphere S^2 by means of a vector field $\mathbf{n}$ (see Fig. 6).

The total curvature K of the second kind is the limit of the quotient of the spherical image area and the area of the original when the original tends to M_0. If we denote the areas of the original and of the spherical image by ΔS and $\Delta\sigma$ respectively then

$$K = \lim \frac{\Delta\sigma}{\Delta S}.$$

Let us find some analytic expressions for K. Suppose, at first, that $\mathbf{n}$ is holonomic. Let (u, v) be the curvilinear coordinates in the surface orthogonal to $\mathbf{n}$ and $\mathbf{x}(u, v)$ be a position vector of that surface, such that $\mathbf{x}_u$, $\mathbf{x}_v$, $\mathbf{n}$ form the positively oriented triple. Denote by $W\,du\,dv$ the area element of the surface. Then

$$\begin{aligned} K = \frac{(\mathbf{n}_u, \mathbf{n}_v, \mathbf{n})}{W} = \frac{1}{W}&(\mathbf{n}_{x_1}, \mathbf{n}_{x_2}, \mathbf{n})\left(\frac{\partial x_1}{\partial u}\frac{\partial x_2}{\partial v} - \frac{\partial x_1}{\partial v}\frac{\partial x_2}{\partial u}\right) \\ &+ (\mathbf{n}_{x_2}, \mathbf{n}_{x_3}, \mathbf{n})\left(\frac{\partial x_2}{\partial u}\frac{\partial x_3}{\partial v} - \frac{\partial x_2}{\partial v}\frac{\partial x_3}{\partial u}\right) \\ &+ (\mathbf{n}_{x_3}, \mathbf{n}_{x_1}, \mathbf{n})\left(\frac{\partial x_3}{\partial u}\frac{\partial x_1}{\partial v} - \frac{\partial x_3}{\partial v}\frac{\partial x_1}{\partial u}\right). \end{aligned}$$

Note that $\mathbf{n}$ can be represented as

$$\mathbf{n} = \frac{[\mathbf{x}_u, \mathbf{x}_v]}{W}.$$

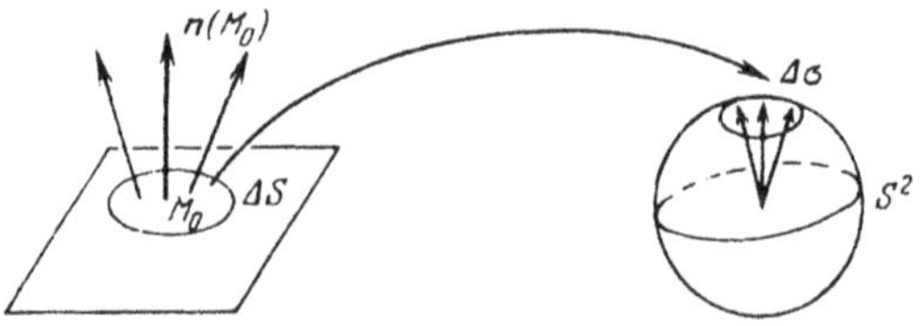

FIGURE 6

Introduce a vector **P** setting

$$\mathbf{P} = \{(\mathbf{n}_{x_2}, \mathbf{n}_{x_3}, \mathbf{n}), (\mathbf{n}_{x_3}, \mathbf{n}_{x_1}, \mathbf{n}), (\mathbf{n}_{x_1}, \mathbf{n}_{x_2}, \mathbf{n})\}.$$

It is easy to show that the definition of **P** is independent of the choice of Cartesian coordinates, i.e. in a coordinate change the components of **P** behave as vector components. So, it is possible to write

$$K = (\mathbf{P}, \mathbf{n}). \tag{1}$$

Now represent K by means of components of the vector which is collinear with **n** but is not in general of unit length. Suppose that such a vector is $\mathbf{A} = \{A_i\} = \lambda\mathbf{n}$, where $\lambda = (A_1^2 + A_2^2 + A_3^2)^{1/2}$. Then evidently

$$\mathbf{P} = \lambda^{-3}\{(A_{x_2}A_{x_3}A), (A_{x_3}A_{x_1}A), (A_{x_1}A_{x_2}A)\}.$$

Therefore, from (1) we see that

$$K = \lambda^{-4}\{(A_{x_2}A_{x_3}A)A_1 + (A_{x_3}A_{x_1}A)A_2 + (A_{x_1}A_{x_2}A)A_3\}$$

$$= -\frac{\begin{vmatrix} A_{1x_1} & A_{1x_2} & A_{1x_3} & A_1 \\ A_{2x_1} & A_{2x_2} & A_{2x_3} & A_2 \\ A_{3x_1} & A_{3x_2} & A_{3x_3} & A_3 \\ A_1 & A_2 & A_3 & 0 \end{vmatrix}}{(A_1^2 + A_2^2 + A_3^2)^2}. \tag{2}$$

Let us be given a family of surfaces $\Phi(x_1, x_2, x_3) = \text{const}$, which are orthogonal to **n**. Then we may take $\operatorname{grad}\Phi$ as **A** and obtain the Neumann formula

$$K = -\frac{\begin{vmatrix} \Phi_{x_1x_1} & \Phi_{x_1x_2} & \Phi_{x_1x_3} & \Phi_{x_1} \\ \Phi_{x_1x_2} & \Phi_{x_2x_2} & \Phi_{x_2x_3} & \Phi_{x_2} \\ \Phi_{x_1x_3} & \Phi_{x_2x_3} & \Phi_{x_3x_3} & \Phi_{x_3} \\ \Phi_{x_1} & \Phi_{x_2} & \Phi_{x_3} & 0 \end{vmatrix}}{(\Phi_{x_1}^2 + \Phi_{x_2}^2 + \Phi_{x_3}^2)^2}. \tag{3}$$

Turn now to the general case, when the field **n** may be non-holonomic. We shall show that (1) holds in this case, too. At first, clarify the geometrical meaning of **P**. Let F be an arbitrary surface located in the domain of definition of the field **n**. Let the parameters (α, β) in the surface induce the normal vector field ν. We can map the surface F into the unit sphere S^2 by putting each point $M \in F$ into correspondence with the end-point of $\mathbf{n}(M)$ in S^2. Denote this mapping by ψ. The area element $d\sigma$ of unit sphere can be represented as

$$d\sigma = (\mathbf{n}_\alpha, \mathbf{n}_\beta, \mathbf{n})\, d\alpha\, d\beta. \tag{4}$$

Let $x_i = x_i(\alpha, \beta)$ be the components of the surface F position vector with respect to Cartesian coordinates. Then we have

$$(\mathbf{n}_\alpha, \mathbf{n}_\beta, \mathbf{n}) = \left\{ (\mathbf{n}_{x_2}, \mathbf{n}_{x_3}, \mathbf{n}) \left(\frac{\partial x_2}{\partial \alpha} \frac{\partial x_3}{\partial \beta} - \frac{\partial x_2}{\partial \beta} \frac{\partial x_3}{\partial \alpha} \right) \right.$$
$$+ (\mathbf{n}_{x_3}, \mathbf{n}_{x_1}, \mathbf{n}) \left(\frac{\partial x_3}{\partial \alpha} \frac{\partial x_1}{\partial \beta} - \frac{\partial x_3}{\partial \beta} \frac{\partial x_1}{\partial \alpha} \right)$$
$$\left. + (\mathbf{n}_{x_1}, \mathbf{n}_{x_2}, \mathbf{n}) \left(\frac{\partial x_1}{\partial \alpha} \frac{\partial x_2}{\partial \beta} - \frac{\partial x_1}{\partial \beta} \frac{\partial x_2}{\partial \alpha} \right) \right\}. \tag{5}$$

Let ν_i be the components of surface F unit normal, dS be the area element of F. From (4) and (5) we obtain

$$d\sigma = (\mathbf{P}, \nu)\, dS. \tag{6}$$

Consider a neighborhood of some point M in F. Let ΔS and $\Delta\sigma$ be the areas of that neighborhood and its spherical image respectively. Then (6) shows that

$$\lim \frac{\Delta\sigma}{\Delta S} = (\mathbf{P}, \nu). \tag{7}$$

where ν is the normal to F at M. Thus, *there is a vector field* **P** *depending on* **n** *such that for any unit vector* ν *the transversal Jacobian of the mapping* ψ *of the plane orthogonal to* ν *into the unit sphere* S^2 *is equal to* $(\mathbf{P}, \nu)$. *We call* **P** *the vector of curvature of the field* **n**.

The total curvature K can be introduced in another way. Consider the Rodrigues equation $d\mathbf{n} = -\lambda\, dx$. In general, this equation is insolvable for arbitrary field **n**, excluding the trivial case $\lambda = 0$. To find λ we have the equation

$$\det \|\xi_{ix_j} + \lambda\delta_{ij}\| = 0.$$

Since $\det \|\xi_{ix_j}\|$ is the Jacobian of the mapping of a three-dimensional domain in E^3 into the unit two-dimensional sphere S^2, $\det \|\xi_{ix_j}\| = 0$. Hence $\lambda = 0$ is a root. The other two roots satisfy the characteristic equation

$$\lambda^2 + \lambda(\xi_{1x_1} + \xi_{2x_2} + \xi_{3x_3}) + \frac{1}{2} \sum_{i,j=1}^{3} \begin{vmatrix} \xi_{ix_i} & \xi_{ix_j} \\ \xi_{jx_i} & \xi_{jx_j} \end{vmatrix} = 0$$

We call the roots of the above equation the *principal curvatures of the second kind*. In spite of the fact that the roots can be complex when the equation $d\mathbf{n} = -\lambda\, d\mathbf{x}$ is insolvable, the half-sum of roots is called, as usual, the mean curvature H and the product of roots is called the total curvature K:

$$H = -\frac{1}{2} \operatorname{div} \mathbf{n} = \frac{1}{2}(\lambda_1 + \lambda_2),$$
$$K = \lambda_1 \lambda_2 = \frac{1}{2} \sum_{i,j=1}^{3} \begin{vmatrix} \xi_{ix_i} & \xi_{ix_j} \\ \xi_{jx_i} & \xi_{jx_j} \end{vmatrix}. \tag{8}$$

Indeed, since the expression for $\lambda_1\lambda_2$ is invariant with respect to the choice of Cartesian coordinates, we shall prove (8) with respect to the special system of coordinates. Directing the axis x_3 along $\mathbf{n}(M)$, we get

$$\lambda_1\lambda_2 = \begin{vmatrix} \xi_{1x_1} & \xi_{1x_2} \\ \xi_{2x_1} & \xi_{2x_2} \end{vmatrix}. \qquad (9)$$

On the other hand, we have (1). Taking into account that

$$\mathbf{P} = \left\{ \begin{vmatrix} \xi_{1x_2} & \xi_{1x_3} \\ \xi_{2x_2} & \xi_{2x_3} \end{vmatrix}, \begin{vmatrix} \xi_{1x_3} & \xi_{1x_1} \\ \xi_{2x_3} & \xi_{2x_1} \end{vmatrix}, \begin{vmatrix} \xi_{1x_1} & \xi_{1x_2} \\ \xi_{2x_1} & \xi_{2x_2} \end{vmatrix} \right\}.$$

and the fact that the third component of $\mathbf{P}$ is its projection onto $\mathbf{n}$, we find that $K = \lambda_1\lambda_2$ for any vector field $\mathbf{n}$.

Comparing (6), (7) of Section 1.2 with (8) and (9), we see that *the mean curvatures of the first and second kinds coincide, while the total curvatures of the first and second kinds are different in the non-holonomic case:* $K - K_1 = \frac{(\mathbf{n},\operatorname{curl}\mathbf{n})^2}{4}$.

In the theory of surfaces it is well known that $H^2 \geq K$. *The generalization of this inequality to the case of an arbitrary vector field is the following:*

$$H^2 + \frac{(\mathbf{n},\operatorname{curl}\mathbf{n})^2}{4} \geq K. \qquad (10)$$

We may state (10) in the special system of coordinates constructed above with respect to which $\xi_1 = \xi_2 = \xi_{3x_3} = 0$. In this case we have

$$H = -\frac{1}{2}(\xi_{1x_1} + \xi_{2x_2}), \qquad (\mathbf{n},\operatorname{curl}\mathbf{n}) = \xi_{2x_1} - \xi_{1x_2},$$

$$K = \begin{vmatrix} \xi_{1x_1} & \xi_{1x_2} \\ \xi_{2x_1} & \xi_{2x_2} \end{vmatrix},$$

$$H^2 + \frac{(\mathbf{n},\operatorname{curl}\mathbf{n})^2}{4} - K = \left(\frac{\xi_{1x_1} - \xi_{2x_2}}{2}\right)^2 + \left(\frac{\xi_{1x_2} + \xi_{2x_1}}{2}\right)^2 \geq 0.$$

We call the curvatures H, K, the value of non-holonomicity $(\mathbf{n},\operatorname{curl}\mathbf{n})$ and the curvature of streamlines $|k|$ *the basic invariants of the field*. Since $\lambda_i = H \pm \sqrt{H^2 - K}$, from (10) we obtain the estimate of the imaginary part of the root of the characteristic equation

$$|\operatorname{Im}\lambda_i| \leq \sqrt{K - H^2} \leq \left|\frac{(\mathbf{n},\operatorname{curl}\mathbf{n})}{2}\right|.$$

Let λ_1, λ_2 be the roots of equation $d\mathbf{n} = -\lambda\, d\mathbf{x}$, where the shift $d\mathbf{x}$ is orthogonal to $\mathbf{n}$. The following theorem holds.

Theorem *If at each point* $\lambda_1 = \lambda_2 \neq 0$ *then the field* $\mathbf{n}$ *is holonomic and the surfaces orthogonal to* $\mathbf{n}$ *are spheres.*

Denote by λ the common value of λ_i. It is real. Let τ, ν be the fields orthogonal to $\mathbf{n}$. As $\lambda_1 = \lambda_2$, an arbitrary direction which is orthogonal to $\mathbf{n}$, satisfies the equation

$$d\mathbf{n} = -\lambda\, d\mathbf{x}.$$

We have

$$\mathbf{n}_{x_i}\tau_i = -\lambda\tau, \qquad \mathbf{n}_{x_j}\nu_j = -\lambda\nu.$$

Differentiating the first equation via ν and the second via τ and then subtracting, we get

$$(\mathbf{n}_{x_i}\tau_i)_{x_j}\nu_j - (\mathbf{n}_{x_j}\nu_j)_{x_i}\tau_i = (-\lambda\tau)_{x_j}\nu_j + (\lambda\nu)_{x_i}\tau_i.$$

The latter expression can be rewritten as

$$\mathbf{n}_{x_i}(\tau_{ix_j}\nu_j - \nu_{ix_j}\tau_j) = -\lambda(\nabla_\nu\tau - \nabla_\tau\nu) - [[\nu,\tau], \operatorname{grad}\lambda].$$

If we multiply both sides scalarly by $\mathbf{n}$ then we obtain zero on the left-hand side due to the fact that $\mathbf{n}$ is of unit length and as a consequence $(\mathbf{n}, \mathbf{n}_{x_i}) = 0$. Therefore

$$(\mathbf{n}, \nabla_\nu\tau - \nabla_\tau\nu) = 0.$$

In Section 1.1 we stated that this means holonomicity of the vector field $\mathbf{n}$. Since the surface which is orthogonal to $\mathbf{n}$ is totally umbilic, it is a sphere by the well-known theorem from the theory of surfaces.

As an exercise, find the total curvature of the second kind for the special field $\mathbf{n}$ whose streamlines are the family of small circles in concentric spheres. Let O be a common center of the spheres. Suppose that the centers of the circles — the streamlines in the fixed sphere — are on the same diameter l. Also, we suppose that the circles are in the parallel planes orthogonal to l. At two opposite points of the sphere the circles degenerate into points — the singular points of the vector field $\mathbf{n}$. The diameter l is a function of the sphere radius. We consider the simplest case, when the diameter is in the x, y-plane. Denote by α the angle which l makes with the axis Ox. If x, y, z are the coordinates of some point on a sphere then $\alpha = \alpha(x^2 + y^2 + z^2)$. In each sphere of the family under consideration the circles make no intersections with each other. Moreover, since the spheres are concentric, the circles of different spheres have no intersections. The tangent vector field $\mathbf{n}$ of the circles is regular at all points excluding the points of some curve which lies in the x, y-plane and is symmetric with respect to O. Let us find the components of $\mathbf{n} = \{\xi_i\}$ as functions of the coordinates. Let the diameter l coincide with the axis Ox. Then at a point with coordinates x, y, z we have

$$\xi_1 = 0, \qquad \xi_2 = -\frac{z}{\sqrt{z^2 + y^2}}, \qquad \xi_3 = \frac{y}{\sqrt{z^2 + y^2}}.$$

Let l make an angle α with Ox. If we rotate the coordinate axes by angle α in such a way that the axis Ox coincides with l then we obtain the previous situation.

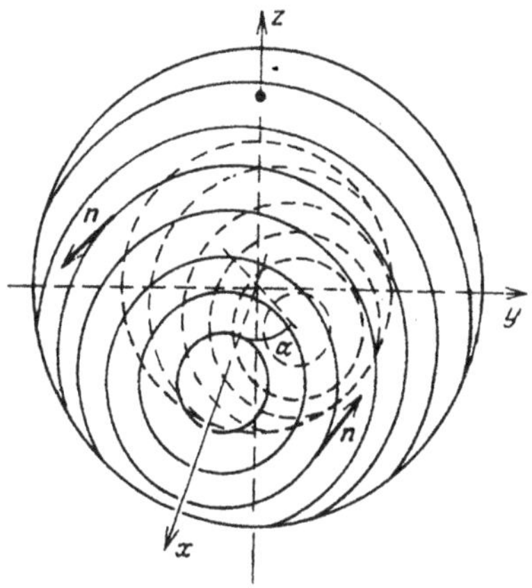

FIGURE 7

The new coordinates u, v, w are related to the old coordinates by the following formulas:

$$u = \cos\alpha x + \sin\alpha y,$$
$$v = -\sin\alpha x + \cos\alpha y,$$
$$w = z.$$

Let η_1, η_2, η_3 be the new components of the vector field **n** on the sphere which corresponds to the diameter l. They are of the form stated above:

$$\eta_1 = 0, \qquad \eta_2 = -\frac{w}{\sqrt{w^2 + v^2}}, \qquad \eta_3 = \frac{v}{\sqrt{w^2 + v^2}}.$$

The old components of **n** have the following expression in terms of the new ones:

$$\xi_1 = \eta_1 \cos\alpha - \eta_2 \sin\alpha,$$
$$\xi_2 = \eta_1 \sin\alpha + \eta_2 \cos\alpha,$$
$$\xi_3 = \eta_3.$$

Therefore, the representative of the vector field **n** at a point with coordinates x, y, z has components

$$\xi_1 = z \sin\alpha/\lambda,$$
$$\xi_2 = -z \cos\alpha/\lambda,$$
$$\xi_3 = (-\sin\alpha x + \cos\alpha y)/\lambda,$$

where $\alpha = \alpha(x^2 + y^2 + z^2)$ is a given function. $\lambda = \sqrt{z^2 + (-\sin\alpha x + \cos\alpha y)^2}$. To find the total curvature K we use formula (2). Set $\mathbf{A} = (z\sin\alpha, -z\cos\alpha, -\sin\alpha x + \cos\alpha y)$. Write the numerator of formula (2)

$$\Delta = \begin{vmatrix} z\cos\alpha\alpha_x & z\cos\alpha\alpha_y & \sin\alpha + z\cos\alpha\alpha_z & z\sin\alpha \\ z\sin\alpha\alpha_x & z\sin\alpha\alpha_y & -\cos\alpha + z\sin\alpha\alpha_z & -z\cos\alpha \\ -\sin\alpha - u\alpha_x & \cos\alpha - u\alpha_y & -u\alpha_z & v \\ z\sin\alpha & -z\cos\alpha & v & 0 \end{vmatrix}.$$

We multiply the elements of the first row by x/z, the second one by y/z and then add to the third one. We multiply the elements of the fourth row by $1/z$ and add to the third one. As a result we obtain a row of zeros. Therefore, $K \equiv 0$. To find the mean curvature of the field we use the formulas

$$-2H = \operatorname{div}\mathbf{n} = \operatorname{div}\frac{\mathbf{A}}{\lambda} = \frac{\operatorname{div}\mathbf{A}}{\lambda} - \frac{(\mathbf{A}, \operatorname{grad}\lambda)}{\lambda^2}.$$

It is easy to see that $\operatorname{div}\mathbf{A} = 0$. Write the expression for the components of $\operatorname{grad}\lambda$:

$$\lambda_x = \frac{v}{\lambda}(-\sin\alpha - u\alpha_x),$$

$$\lambda_y = \frac{v}{\lambda}(\cos\alpha - u\alpha_y),$$

$$\lambda_z = \frac{1}{\lambda}(z - uv\alpha_z).$$

The direct calculation gives $(\mathbf{A}, \operatorname{grad}\lambda) = 0$. Therefore, $H \equiv 0$. It is easy to calculate the non-holonomicity of the vector field just constructed. We have

$$\lambda^2(\mathbf{n}, \operatorname{curl}\mathbf{n}) = (\mathbf{A}, \operatorname{curl}\mathbf{A}) = -2z(x^2 + y^2 + z^2)\alpha'.$$

Therefore, if $\alpha' \neq 0$ and $z \neq 0$ then $\mathbf{n}$ is non-holonomic. Thus, for the vector field under consideration we have $K = H \equiv 0$ and $(\mathbf{n}, \operatorname{curl}\mathbf{n}) \neq 0$.

Now we give the geometric meaning of the field $\mathbf{P}$. *If* $\mathbf{P} \neq 0$ *at some point* M *in the domain of definition of the field* $\mathbf{n}$ *then a unique line* $\mathbf{n} = \text{const}$, *which is tangent to* $\mathbf{P}$ *at each of its points, passes through that point.*

Indeed, suppose that $d\mathbf{r} = (dx_1, dx_2, dx_3)$ is tangent to the line $\mathbf{n} = \text{const}$. Choosing the special system of coordinates as above, we find that $d\mathbf{r}$ satisfies the following equations at M:

$$\xi_{1x_1}\,dx_1 + \xi_{1x_2}\,dx_2 + \xi_{1x_3}\,dx_3 = 0,$$
$$\xi_{2x_1}\,dx_1 + \xi_{2x_2}\,dx_2 + \xi_{2x_3}\,dx_3 = 0.$$

From this we find that $d\mathbf{r} = \lambda\mathbf{P}$, where λ is some number. Therefore, the line $\mathbf{n} = \text{const}$ is tangent to $\mathbf{P}$.

From the formula $K = (\mathbf{n}, \mathbf{P})$ it follows *a simple way to construct the field* $\mathbf{n}$ *with zero total curvature*: $K = 0$. Indeed, that condition means that $\mathbf{n}$ and $\mathbf{P}$ are mutually orthogonal. Suppose that $\mathbf{P} \neq 0$. Since $\mathbf{n}$ is constant along the streamline of $\mathbf{P}$, these streamlines are planar if $K = 0$. Therefore, to construct the required field it is sufficient to define an arbitrary congruence of planar curves and set $\mathbf{n} = \text{const}$ on each curve of the family. To do this, take a surface F^2 of the position vector $\mathbf{r}(u, v)$. Then at each point $x \in F^2$ take a unit vector $\mathbf{n}(u, v)$ and a vector $\mathbf{a}(x)$ in a plane which is orthogonal to $\mathbf{n}(x)$. We realize that selection with the help of a single function $\varphi(u, v)$. We emit a curve in the direction of $\mathbf{a}(x)$ which lies in the plane orthogonal to $\mathbf{n}(x)$. Those curves are defined by curvature $k(u, v, s)$ as a function of the arc length parameter. So, the arbitrariness in the definition of $\mathbf{n}$ is in six functions of two arguments and one function of three arguments.

In the case of $\mathbf{P} = 0$, the image of any surface generated on the unit sphere by the vector field $\mathbf{n}$ is a line. Conversely, the inverse image of any point of that line is a surface.

To construct the field $\mathbf{n}$ satisfying $\mathbf{P} = 0$ define a regular family of surfaces in E^3 and then define some constant value of $\mathbf{n}$ on each of the surfaces which varies continuously on the family.

1.6 The Asymptotic Lines

We call a direction $d\mathbf{x}$ in a plane orthogonal to the field $\mathbf{n}$ *asymptotic* if the normal curvature of the field in the direction $d\mathbf{x}$ is zero. Since the normal curvature of the field is equal to the quotient of $-(d\mathbf{n}, d\mathbf{x})$ and $d\mathbf{x}^2$, the asymptotic direction is orthogonal to $d\mathbf{n}$. Therefore, $d\mathbf{x}$ is proportional to $[\mathbf{n}, d\mathbf{n}]$. Thus, the asymptotic direction can be found from the following vector equation:

$$[\mathbf{n}, d\mathbf{n}] = \mu\, d\mathbf{x},$$

where μ is some numeric coefficient. Find the condition when the asymptotic direction exists. To do this we use a special system of coordinates such that $\xi_1 = \xi_2 = 0$, $\xi_3 = 1$ at some fixed point P_0. Then at P_0 the system on the asymptotic direction takes the form

$$\begin{aligned} -(\xi_{2x_1} + \mu)\, dx_1 - \xi_{2x_2}\, dx_2 &= 0, \\ \xi_{1x_1}\, dx_1 + (\xi_{1x_2} - \mu)\, dx_2 &= 0, \\ dx_3 &= 0. \end{aligned}$$

Therefore, the coefficient μ satisfies the quadratic equation

$$\mu^2 - \mu(\xi_{1x_2} - \xi_{2x_1}) + (\xi_{1x_1}\xi_{2x_2} - \xi_{1x_2}\xi_{2x_1}) = 0.$$

Rewrite this equation in terms of the field invariants:

$$\mu^2 - \mu(\mathbf{n}, \operatorname{curl} \mathbf{n}) + K = 0.$$

We have

$$\mu_{1,2} = \frac{(\mathbf{n}, \operatorname{curl} \mathbf{n}) \pm \sqrt{(\mathbf{n}, \operatorname{curl} \mathbf{n})^2 - 4K}}{2}.$$

Note that $-4K_1 = (\mathbf{n}, \operatorname{curl}\mathbf{n})^2 - 4K$. Therefore, the asymptotic direction exists if and only if the total curvature of the first kind satisfies $K_1 \leq 0$. *If $K_1 < 0$ then there are two asymptotic directions. If $K_1 = 0$ then either only one direction is asymptotic or any direction orthogonal to* $\mathbf{n}$ *is asymptotic.*

The line which is tangent to the asymptotic direction at each of its points is called *the asymptotic line*. Let us clarify the geometrical meaning of μ. Let s be the arc length parameter of an asymptotic line. Then along the asymptotic line we have

$$\frac{d\mu}{ds}\frac{d\mathbf{r}}{ds} + \mu\frac{d^2\mathbf{r}}{ds^2} = \left[\mathbf{n}, \frac{d^2\mathbf{n}}{ds^2}\right].$$

Multiplying by $\mathbf{n}$ we have $\left(\frac{d^2\mathbf{r}}{ds^2}, \mathbf{n}\right) = 0$, provided that $\mu \neq 0$. Therefore, the vector $\mathbf{n}$ is a binormal of the asymptotic line or differs from it by sign. Denote a torsion of the asymptotic line by κ and the principal normal and binormal vectors by ν and β respectively. If $\mathbf{n} = \beta$ then $d\mathbf{n}/ds = -\kappa\nu$. We have

$$\mu\frac{d\mathbf{r}}{ds} = \left[\mathbf{n}, \frac{d\mathbf{n}}{ds}\right] = -\kappa[\mathbf{n}, \nu] = -\kappa\frac{d\mathbf{r}}{ds}.$$

Therefore, the value of μ differs from κ only in its sign. The same result occurs when $\mathbf{n} = -\beta$. Thus, *the torsion of the asymptotic line can be expressed in terms of the field basic invariants as*

$$\kappa = \frac{-\left((\mathbf{n}, \operatorname{curl}\mathbf{n}) \pm \sqrt{(\mathbf{n}, \operatorname{curl}\mathbf{n})^2 - 4K}\right)}{2}.$$

Denote the torsion of the first and the second asymptotic lines by κ_i, $i = 1, 2$, respectively. The following generalization of the Beltrami–Enneper theorem holds:

Theorem *The sum of the torsions of the asymptotic lines is equal to the field non-holonomicity with a negative sign and the product of the torsions of the asymptotic lines is equal to the total curvature of the second kind:*

$$\kappa_1 + \kappa_2 = -(\mathbf{n}, \operatorname{curl}\mathbf{n}), \quad \kappa_1\kappa_2 = K.$$

If $K = 0$ then the torsion of one of the asymptotic lines is zero and the torsion of the other is equal to the field non-holonomicity value up to a sign.

As an example, find the asymptotic lines of the helical field, i.e. the field of helical streamlines. Consider the regular family of helices on the cylinders with a common axis. Suppose that the helices on the same cylinder have the same pitch of screw. We suppose, also, that the pitch of a screw depends on the cylinder radius. Consider the equation of some of these helices:

$$x = \rho\cos\varphi,$$
$$y = \rho\sin\varphi,$$
$$z = b\varphi,$$

where $\rho = \sqrt{x^2 + y^2}$, $b = b(\rho^2)$ is some function of ρ^2. The helix tangent vector is

$$A = \begin{pmatrix} x_\varphi \\ y_\varphi \\ z_\varphi \end{pmatrix} = \begin{pmatrix} -y \\ x \\ b \end{pmatrix}.$$

Also, consider the unit vector field

$$\mathbf{n} = \left\{ \frac{-y}{\sqrt{\rho^2 + b^2}}, \frac{x}{\sqrt{\rho^2 + b^2}}, \frac{b}{\sqrt{\rho^2 + b^2}} \right\}.$$

The field does not depend on z. If $b(0) \neq 0$ then the field is well defined for $x = 0$, $y = 0$, also. Find the mean curvature of the field

$$2H = -\operatorname{div} \mathbf{n} = \frac{\partial}{\partial x} \frac{y}{\sqrt{\rho^2 + b^2}} - \frac{\partial}{\partial y} \frac{x}{\sqrt{\rho^2 + b^2}} = 0.$$

We find the total curvature of the second kind by (2) of Section 1.5:

$$K = -\frac{1}{\sqrt{\rho^2 + b^2}} \begin{vmatrix} 0 & -1 & 0 & -y \\ 1 & 0 & 0 & x \\ b'2x & b'2y & 0 & b \\ -y & x & b & 0 \end{vmatrix} = \frac{-b(2\rho^2 b' - b)}{(\rho^2 + b^2)^2}.$$

It is easy to see that $(\mathbf{A}, \operatorname{curl} \mathbf{A}) = 2(b'\rho^2 - b)$. Therefore, the field non-holonomicity value is

$$(\mathbf{n}, \operatorname{curl} \mathbf{n}) = \frac{2(b'\rho^2 - b)}{(\rho^2 + b^2)^2}.$$

From this we see that $\mathbf{n}$ is holonomic if $b = c\rho^2$, where c is a constant. Write the equation of asymptotic lines in coordinate form:

$$\xi_2\, d\xi_3 - \xi_3\, d\xi_2 = \mu\, dx,$$
$$\xi_3\, d\xi_1 - \xi_1\, d\xi_3 = \mu\, dy,$$
$$\xi_1\, d\xi_2 - \xi_2\, d\xi_1 = \mu\, dz.$$

Represent the field components in terms of ρ and φ:

$$\xi_1 = \frac{-\rho \sin\varphi}{\sqrt{\rho^2 + b^2}}, \qquad \xi_2 = \frac{\rho \cos\varphi}{\sqrt{\rho^2 + b^2}}, \qquad \xi_3 = \frac{b}{\sqrt{\rho^2 + b^2}}.$$

The equations of asymptotic lines can then be represented as

$$\begin{aligned} (\rho \cos\varphi db - bd(\rho \cos\varphi)) &= (\rho^2 + b^2)\mu d(\rho \cos\varphi), \\ (\rho \sin\varphi db - bd(\rho \sin\varphi)) &= (\rho^2 + b^2)\mu d(\rho \sin\varphi), \\ \rho^2\, d\varphi &= (\rho^2 + b^2)\mu\, dz. \end{aligned} \tag{1}$$

For the subradical expression in the formula for evaluating μ at the beginning of this section we have

$$(\mathbf{n}, \operatorname{curl} \mathbf{n})^2 - 4K = \frac{4b'^2\rho^4}{(\rho^2 + b^2)^2}.$$

Therefore, the helical field has two, maybe coinciding, asymptotic directions. The coefficient μ in the equation of asymptotic lines takes one of two values:

$$\mu_1 = \frac{-b}{\rho^2 + b^2}, \qquad \mu_2 = \frac{2b'\rho^2 - b}{\rho^2 + b^2}.$$

If $\mu = \mu_1 = -b/(\rho^2 + b^2)$ then from the first and second equations of (1) it follows that $db = 0$, while from the third we see that $\rho^2\, d\varphi = -b\, dz$. Therefore, in this case the asymptotic line is on the cylinder $\rho = \text{const}$ and represents a helix with torsion of the sign opposite to the sign of torsion of the field streamline. The family of these helices together with the original family of helices form a regular orthogonal net on each cylinder. Consider the asymptotic lines corresponding to $\mu = \mu_2$. The equations of those asymptotic lines have the form

$$\rho \cos\varphi\, db = 2b'\rho^2 d(\rho \cos\varphi),$$
$$\rho \sin\varphi\, db = 2b'\rho^2 d(\rho \sin\varphi),$$
$$\rho^2\, d\varphi = (2b'\rho^2 - b)\, dz.$$

Provided that $b' \neq 0$, it follows from the first and the second equations that $d\varphi = 0$. From the third equation we see that $dz = 0$, provided that $K \neq 0$. Therefore, in this case asymptotic lines are the rays emitted parallel to the x, y-plane from the points of the common axis of the cylinders. This solution is evident from a descriptive viewpoint. In this case we ought to regard the torsion κ_2 as a value which characterizes the rotation of the vector field $\mathbf{n}$ in moving along that ray.

If $b' = 0$ then $K_1 = 0$ and $\mu_1 = \mu_2$. In this case the first and second equations in (1) turn into identities. The third equation $\rho^2\, d\varphi = -b\, dz$ expresses the orthogonality condition of the required asymptotic direction and given field $\mathbf{n}$. Therefore, in the case of $b' = 0$ any direction which is orthogonal to $\mathbf{n}$ is asymptotic.

1.7 The First Divergent Form of Total Curvature of the Second Kind

Now we are going to state an integral formula which gives the expression for the integral of total curvature K in a domain G in terms of some integral along the boundary ∂G or in terms of an integral over some domain on the unit sphere S^2. Let $\mathbf{x}$ be a position vector of $M \in G$. Suppose that (α, β) are local curvilinear coordinates in ∂G with the usual orientation, ψ is a mapping of ∂G into the unit sphere S^2 by means of the vector field $\mathbf{n}$, $d\sigma = (\mathbf{n}_\alpha, \mathbf{n}_\beta, \mathbf{n})\, d\alpha\, d\beta$ is a signed area element of unit sphere. *Then supposing no singular points of the vector field* **n** *in* G, *the following formula holds:*

$$\int_G K\,dV = \int_{\nu(\partial G)} (\mathbf{x}, \mathbf{n})\, d\sigma, \tag{1}$$

where dV *is the volume element of a three-dimensional domain* G.

Proof We can represent the integrand on the right as

$$(\mathbf{x}, \mathbf{n})\, d\sigma = (\mathbf{x}, \mathbf{n})(\mathbf{n}_\alpha, \mathbf{n}_\beta, \mathbf{n})\, d\alpha\, d\beta$$

$$= (\mathbf{x}, \mathbf{n})\left\{(\mathbf{n}_{x_1}, \mathbf{n}_{x_2}, \mathbf{n})\left(\frac{\partial x_1}{\partial \alpha}\frac{\partial x_2}{\partial \beta} - \frac{\partial x_1}{\partial \beta}\frac{\partial x_2}{\partial \alpha}\right)\right.$$

$$+ (\mathbf{n}_{x_2}, \mathbf{n}_{x_3}, \mathbf{n})\left(\frac{\partial x_2}{\partial \alpha}\frac{\partial x_3}{\partial \beta} - \frac{\partial x_2}{\partial \beta}\frac{\partial x_3}{\partial \alpha}\right)$$

$$\left.+ (\mathbf{n}_{x_3}, \mathbf{n}_{x_1}, \mathbf{n})\left(\frac{\partial x_3}{\partial \alpha}\frac{\partial x_1}{\partial \beta} - \frac{\partial x_3}{\partial \beta}\frac{\partial x_1}{\partial \alpha}\right)\right\} d\alpha\, d\beta = (\mathbf{x}, \mathbf{n})(\mathbf{P}, \nu)\, dS.$$

where ν is the normal of ∂G.

By the Gauss–Ostrogradski formula we have

$$\int_{\partial G} (\mathbf{x}, \mathbf{n})(\mathbf{P}, \nu)\, dS = \int_G \operatorname{div}((\mathbf{x}, \mathbf{n})\mathbf{P})\, dV = \int_G (\mathbf{x}, \mathbf{n}) \operatorname{div} \mathbf{P}\, dV + \int_G (\mathbf{P}, \operatorname{grad}(\mathbf{x}, \mathbf{n}))\, dV$$

$$= \int_G (\mathbf{P}, \mathbf{n})\, dV + \int_G \{(\mathbf{x}, \mathbf{n}_{x_1})(\mathbf{n}_{x_2}, \mathbf{n}_{x_3}, \mathbf{n}) + (\mathbf{x}, \mathbf{n}_{x_2})(\mathbf{n}_{x_3}, \mathbf{n}_{x_1}, \mathbf{n})$$

$$+ (\mathbf{x}, \mathbf{n}_{x_3})(\mathbf{n}_{x_1}, \mathbf{n}_{x_2}, \mathbf{n})\}\, dV.$$

The first integral is the integral of total curvature K because of (1) Section 1.5. Since **n** is of unit length, $(\mathbf{n}_{x_1}, \mathbf{n}_{x_2}, \mathbf{n}_{x_3}) = 0$. Consider the coefficient of x_1 in the integrand of the second integral:

$$\xi_{1x_3}(\mathbf{n}_{x_1}, \mathbf{n}_{x_2}, \mathbf{n}) + \xi_{1x_1}(\mathbf{n}_{x_2}, \mathbf{n}_{x_3}, \mathbf{n}) + \xi_{1x_2}(\mathbf{n}_{x_3}, \mathbf{n}_{x_1}, \mathbf{n})$$

$$= -\begin{vmatrix} \xi_{1x_1} & \xi_{1x_2} & \xi_{1x_3} & \xi_1 \\ \xi_{2x_1} & \xi_{2x_2} & \xi_{2x_3} & \xi_2 \\ \xi_{3x_1} & \xi_{3x_2} & \xi_{3x_3} & \xi_3 \\ \xi_{1x_1} & \xi_{1x_2} & \xi_{1x_3} & 0 \end{vmatrix} = (\mathbf{n}_{x_1}, \mathbf{n}_{x_2}, \mathbf{n}_{x_3}) = 0.$$

In an analogous way we see that the coefficients of x_2 and x_3 are both zero. Thus, the second integral turns into zero and, as a consequence, (1) is proved. The calculations above shows that K, as well as the H, is a divergence of some vector. Namely,

$$K = \operatorname{div}(\mathbf{x}, \mathbf{n})\mathbf{P}. \tag{2}$$

1.8 The Second Divergent Representation of Total Curvature of the Second Kind

Consider (8) of Section 1.5 in detail:

$$K = \frac{1}{2}\sum_{i,j=1}^{3}(\xi_{ix_i}\xi_{jx_j} - \xi_{ix_j}\xi_{jx_i})$$
$$= \frac{\partial}{\partial x_1}\frac{1}{2}(\xi_1\xi_{2x_2} - \xi_2\xi_{1x_2}) + \frac{\partial}{\partial x_1}\frac{1}{2}(\xi_1\xi_{3x_3} - \xi_3\xi_{1x_3}) + \cdots,$$

where the dots mean the terms having an analogous structure with respect to replacement of index 1 by 2 or 3. Consider the expression under derivation $\frac{\partial}{\partial x_1}$. It has the form

$$\frac{1}{2}\{\xi_1(\xi_{2x_2} + \xi_{3x_3}) - \xi_2\xi_{1x_2} - \xi_3\xi_{1x_3}\}.$$

Add and substract $\frac{1}{2}\xi_1\xi_{1x_1}$. Then the previous expression obtains the form

$$\frac{1}{2}\{\xi_1(\xi_{1x_1} + \xi_{2x_2} + \xi_{3x_3}) - \xi_1\xi_{1x_1} - \xi_2\xi_{1x_2} - \xi_3\xi_{1x_3}\} = \frac{1}{2}\{\xi_1 \operatorname{div}\mathbf{n} - k_1\},$$

where k_1 is the first component of the field $\mathbf{n}$ streamline curvature vector. Transform in an analogous manner the expressions under the derivations $\frac{\partial}{\partial x_2}$, $\frac{\partial}{\partial x_3}$. We get

$$2K = -\operatorname{div}(2H\mathbf{n} + \mathbf{k}), \tag{1}$$

where $\mathbf{k}$ is the field $\mathbf{n}$ streamline curvature vector.

Now consider a closed surface F containing no interior singular points of the field $\mathbf{n}$. Let ν be the normal of F. Then

$$2\int_G K\,dV = -\int_G (2H\mathbf{n} + \mathbf{k}, \nu)\,dS. \tag{2}$$

Consider some particular cases.

(1) Suppose that *the boundary of G is formed by connected surfaces* F_1 and F_2, where F_2 is inside F_1. Suppose that $\mathbf{n}$ coincides with the normal vector field on F_i (see Fig. 8). In this case $(\mathbf{k}, \nu) = 0$ at the points of surfaces. Since F_i are orthogonal to $\mathbf{n}$, the curvatures of F_i coincide with the curvatures of the field $\mathbf{n}$ at the points of F_i. From (2) it follows that for any vector field $\mathbf{n}$ inside G with F_1 and F_2 as the boundary *the volume integral of field curvature K in G can be expressed as a difference of integrated mean curvatures of boundary surfaces*:

$$\int_G K\,dV = -\int_{F_1} H\,dS + \int_{F_2} H\,dS. \tag{3}$$

Note that the right-hand side is independent of $\mathbf{n}$.

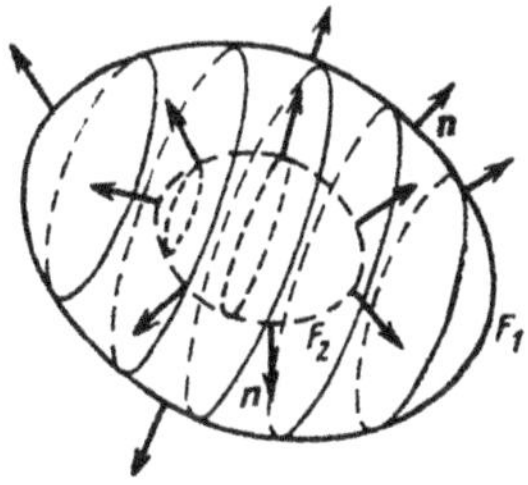

FIGURE 8

(2) Suppose that *at each point of a surface F the field* **n** *is tangent to F*. Then the streamline of **n** starting at F belongs to F. In this case we say that *F is the invariant submanifold of the field* **n**. The value of $(\mathbf{k}, \nu)$ is the normal curvature of the streamline in F, where ν is directed outside F. If **n** is regular at each point of F then F is necessarily homeomorphic to the torus because only the torus admits the regular unit tangent vector field (see Fig. 9). *Formula (2) for this case has the following form:*

$$2\int_G K\,dV = -\int_F k_n\,dS, \tag{4}$$

where G is the interior of F, k_n is the streamline normal curvature in F of the field **n**.

Apply formula (4) to the field defined inside the tubular surface. Denote it by F as above. Let Γ be a closed curve of length l, k and κ be its curvature and torsion respectively. The tubular surface F meets any plane normal to F by the circle of constant radius R with its center in Γ. If $\rho(u)$ is a position vector of Γ and (ξ_1, ξ_2, ξ_3) is a natural frame along Γ, then we can represent a position vector of F as

$$\mathbf{r}(u, v) = \rho(u) + R(\cos v\xi_2 + \sin v\xi_3).$$

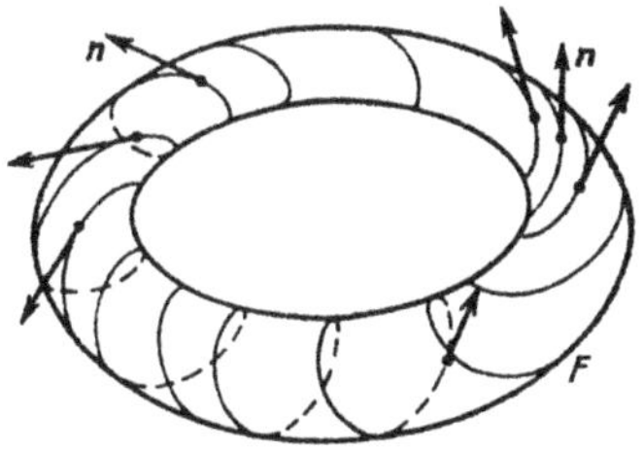

FIGURE 9

We suppose that u is the arc length parameter of Γ. Set

$$\mathbf{a} = \cos v\xi_2 + \sin v\xi_3,$$
$$\mathbf{b} = -\sin v\xi_2 + \cos v\xi_3.$$

Then

$$\mathbf{r}_u = \xi_1(1 - kR\cos v) + R\kappa\mathbf{b}, \qquad \mathbf{r}_v = R\mathbf{b},$$
$$[\mathbf{r}_u, \mathbf{r}_v] = -\mathbf{a}R(1 - kR\cos v).$$

Hence, ξ_1 and $\mathbf{b}$ are tangent to F. The first fundamental form of F has the form

$$ds^2 = \left\{(1 - kR\cos v)^2 + R^2\kappa^2\right\} du^2 + 2R^2\kappa\, du\, dv + R^2\, dv^2.$$

The unit output normal vector ν of F coincides with $\mathbf{a}$.

Let us find the second fundamental form. Since the coefficients of the second fundamental form are equal to the scalar products of the second derivatives of $\mathbf{r}(u, v)$ and the unit normal to F, in second derivatives of $\mathbf{r}(u, v)$ we present only those terms which contain no ξ_1 or b. We have

$$\mathbf{r}_{uu} = k\xi_2(1 - kR\cos v) - R\kappa^2\mathbf{a} + \cdots,$$
$$\mathbf{r}_{uv} = -R\kappa\mathbf{a} + \cdots, \qquad \mathbf{r}_{vv} = -R\mathbf{a}.$$

Hence, the second fundamental form of F with respect to the normal $\mathbf{a}$ has the form

$$II = -\{R\kappa^2 - (1 - kR\cos v)k\cos v\}\, du^2 - 2R\kappa\, du\, dv - R\, dv^2.$$

Since we are interested in the normal curvature of F in the direction of a streamline of $\mathbf{n}$, the differentials du and dv in the expression for I and II must correspond to the directions of the field $\mathbf{n}$ streamline. Introduce the following notations for them: $du = \xi$, $dv = \eta$. The area element of F is

$$dS = R(1 - kR\cos v)\, du\, dv.$$

The following relation holds:

$$II\, dS = \left[kR\cos v I - R^2(\kappa\xi + \eta)^2\right] du\, dv.$$

Hence, the integral of streamline normal curvature over F has the form

$$\int_F k_n\, dS = \int_F \frac{II}{I}\, dS = \int_0^{2\pi}\int_0^{l} Rk\cos v\, du\, dv - \int_0^{2\pi}\int_0^{l} \frac{R^2(\kappa\xi + \eta)^2}{I}\, du\, dv. \tag{5}$$

The first integral on the right is zero because R and k are independent of v. Using (4) and (5) we can write

$$\int_G K\, dV = \int_0^{2\pi}\int_0^{l} \frac{R^2(\kappa\xi + \eta)^2}{2I}\, du\, dv.$$

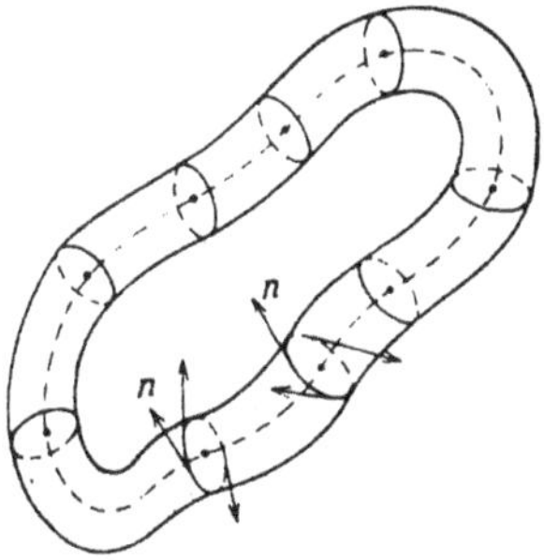

FIGURE 10

As the first fundamental form may be transformed to

$$I = (1 - kR\cos v)^2\xi^2 + R^2(\kappa\xi + \eta)^2$$

the integrand on the right is less then 1/2. So, we have already proved the following statement.

Theorem *If the tubular surface F with the axial line of length l is an invariant surface of the vector field* **n** *then the integral of curvature K of the field* **n** *over a domain bounded by F is non-negative and does not exceed* πl:

$$0 \le \int_G K\,dV \le \pi l.$$

The lowest value of the integral occurs when **n** is such that the streamlines in F satisfy the equation

$$\kappa\xi + \eta = 0$$

(in the particular case, when the torus F axial curve is planar, these curves are $v = \text{const}$).

The largest value corresponds to the field **n** whose streamlines are the circles in planar sections of F perpendicular to the axial curve (see Fig. 10). Note that the upper boundary of the integral does not depend on the tube radius.

1.9 The Interrelation of Two Divergent Representations of the Total Curvatures of the Second Kind

In previous section we stated two divergent representations for K:

$$K = \operatorname{div}(\mathbf{x},\mathbf{n})\mathbf{P}, \qquad 2K = -\operatorname{div}(2H\mathbf{n} + \mathbf{k}).$$

These relations imply formulas (1) of Section 1.7 and (2) of Section 1.8. Comparing them we conclude that if the interior of a domain G contains no singular points of the field $\mathbf{n}$ then

$$\int_{\nu(F)} (\mathbf{x}, \mathbf{n})\, d\sigma = -\int_F \left(H\mathbf{n} + \frac{\mathbf{k}}{2}, \nu\right) dS. \tag{1}$$

We are going to show that (1) is also valid if the singular points are strictly interior to G. Define on F the following external forms:

$$\alpha = 2(\mathbf{x}, \mathbf{n})\, d\sigma, \quad \beta = -\{2H(\mathbf{n}, \nu) + (\mathbf{k}, \nu)\}\, dS, \quad \gamma = (d\mathbf{n}, \mathbf{n}, \mathbf{x}),$$

where $\mathbf{x}$ is a position vector of F, ν is a unit normal of F, dS is an area element of F and $d\sigma$ is an area element of the spherical image of F generated by the vector field $\mathbf{n}$. The definitions of exterior form and exterior differentiation are given in Section 2.9.

Theorem *The difference of β and α is equal to the exterior differential of γ on F:*

$$\beta - \alpha = d\gamma. \tag{2}$$

Let u, v be the curvilinear coordinates in F which induce the normal ν. Then α can be represented as

$$\begin{aligned}\alpha &= 2(\mathbf{x}, \mathbf{n})\, d\sigma = 2(\mathbf{x}, \mathbf{n})(\mathbf{n}_u, \mathbf{n}_v, \mathbf{n})\, du\, dv = 2(\mathbf{n}_u, \mathbf{n}_v, \mathbf{x})\, du\, dv \\ &= \left\{\frac{\partial}{\partial u}(\mathbf{n}, \mathbf{n}_v, \mathbf{x}) + \frac{\partial}{\partial v}(\mathbf{n}_u, \mathbf{n}, \mathbf{x}) + (\mathbf{n}, \mathbf{n}_u, \mathbf{x}_v) - (\mathbf{n}, \mathbf{n}_v, \mathbf{x}_u)\right\} du\, dv \\ &= -d\gamma + (\mathbf{n}, [\mathbf{n}_u, \mathbf{x}_v] - [\mathbf{n}_v, \mathbf{x}_u])\, du\, dv. \end{aligned} \tag{3}$$

Show that the second term on the right is β. Indeed,

$$\begin{aligned}\beta &= \{(\mathbf{n}, \nu)\operatorname{div}\mathbf{n} - (\mathbf{k}, \nu)\}\, |[\mathbf{x}_u, \mathbf{x}_v]|\, du\, dv \\ &= \{(\mathbf{n}, \mathbf{x}_u, \mathbf{x}_v)\operatorname{div}\mathbf{n} - (\mathbf{k}, \mathbf{x}_u, \mathbf{x}_v)\}\, du\, dv.\end{aligned}$$

The forms $(\mathbf{n}, [\mathbf{n}_u, \mathbf{x}_v] - [\mathbf{n}_v, \mathbf{x}_u])\, du\, dv$ and β do not depend on the coordinate systems in E^3 or F. Introduce the coordinates in E^3 and the coordinates u, v in F in such a way that at M_0 the tangent plane to F coincides with the x_1, x_2-coordinate plane and the basis $\mathbf{e}_1 = \mathbf{x}_u$, $\mathbf{e}_2 = \mathbf{x}_v$, $\mathbf{e}_3 = \nu$. Then at M_0

$$\mathbf{n}_u = \mathbf{n}_{x_1}, \qquad \mathbf{n}_v = \mathbf{n}_{x_2}.$$

Moreover, at M_0 we have

$$(\mathbf{n}, [\mathbf{n}_u, \mathbf{x}_v] - [\mathbf{n}_v, \mathbf{x}_u]) = \xi_{1x_1}\xi_3 - \xi_{3x_1}\xi_1 + \xi_{2x_2}\xi_3 - \xi_{3x_2}\xi_2, \tag{4}$$

$$(\mathbf{n}, \mathbf{x}_u, \mathbf{x}_v)\operatorname{div}\mathbf{n} - (\mathbf{k}, \mathbf{x}_u, \mathbf{x}_v) = \xi_{1x_1}\xi_3 + \xi_{2x_2}\xi_3 - \xi_{3x_1}\xi_1 - \xi_{3x_2}\xi_2. \tag{5}$$

Comparing (4) and (5), we conclude

$$(\mathbf{n}, [\mathbf{n}_u, \mathbf{x}_v] - [\mathbf{n}_v, \mathbf{x}_u]) = (\mathbf{n}, \mathbf{x}_u, \mathbf{x}_v)\operatorname{div}\mathbf{n} - (\mathbf{k}, \mathbf{x}_u, \mathbf{x}_v).$$

From (3) it follows that

$$\alpha = -d\gamma + \beta.$$

So, (2) is proved.

Since F is a closed surface,

$$\int_F d\gamma = 0.$$

Hence, (2) implies (1), while the singular points of the field $\mathbf{n}$ possibly exist strictly inside of the domain bounded by F.

Observe that if $\mathbf{n}$ is the field of normals of F the equation (1) implies the *Minkowski theorem on the integrated support function of a closed surface and on its integrated mean curvature*, i.e. (1) turns into the following equality:

$$\int_F (\mathbf{x}, \mathbf{n}) K \, dS = -\int_F H \, dS.$$

Thus, (1) generalizes the Minkowski formula.

1.10 The Generalization of the Gauss–Bonnet Formula for the Closed Surface

In the theory of surfaces the remarkable relation between the integrated Gaussian curvature over some domain G in a surface F and the integrated geodesic curvature of the boundary of G is well known. It is the so-called Gauss–Bonnet formula:

$$\int_G K \, dS = \int_{\partial G} \frac{1}{\rho_g} ds + 2\pi. \tag{1}$$

Here ∂G is a smooth closed curve on a surface F. It is also supposed that the boundary normal ν, with respect to which one evaluates $1/\rho_g$, is directed outward from G. If the surface F is closed then

$$\int_F K \, dS = 4\pi\chi, \tag{2}$$

where χ is an integer called the Euler characteristic of F. We are going to generalize these formulas to vector fields.

Suppose that a vector field $\mathbf{n}$ is defined, maybe with singularities, in a domain Q of three-dimensional Euclidean space. Let F be a closed surface in Q not passing through the singular points.

Theorem *For any closed surface $F \subset Q$ not passing through the singular points of the vector field* $\mathbf{n}$

$$\int_F (\mathbf{n}K + 2H\mathbf{k} + \nabla_{\mathbf{k}}\mathbf{n}, \nu) \, dS = 4\pi\theta, \tag{3}$$

where θ is an integer called the degree of mapping of F into S^2 generated by the vector field **n**.

Here $\nabla_{\mathbf{k}}\mathbf{n}$ means the derivative of **n** along the curvature vector **k** of the **n** field streamline. In the case when **n** is holonomic and F is a surface orthogonal to **n**, i.e. $\mathbf{n} = \nu$, then $(\mathbf{k}, \nu) = 0$, $(\nabla_{\mathbf{k}}\mathbf{n}, \nu) = 0$ and the integral in (3) coincides with the integral of the Gaussian curvature of F. To prove (3) we consider the mapping of F into the unit sphere S^2 generated by the vector field **n**. In Section 1.5 we denoted that mapping by ψ and found that the area elements of image and original are related to each other by the following formula:

$$d\sigma = (\mathbf{P}, \nu)\, dS,$$

where **P** is some invariant vector defined by **n**. In the mapping ψ the image covers S^2 an integer number times. We recall the notion from topology of the mapping degree of a closed surface F onto the unit sphere. Denote it by θ. The mapping degree is connected with ψ image area by the following formula:

$$\int_{\psi(F)} d\sigma = 4\pi\theta.$$

Hence

$$\int_F (\mathbf{P}, \nu)\, dS = 4\pi\theta. \tag{5}$$

Consider the vector **P** in detail. Choose the coordinate axes in such a way that the x_3-axis at $M \in F$ is directed along $\mathbf{n}(M)$. Then the mean and the total curvatures at that point obtain the following form:

$$H = -\frac{1}{2}(\xi_{1x_1} + \xi_{2x_2}), \qquad K = (\xi_{1x_1}\xi_{2x_2} - \xi_{1x_2}\xi_{2x_1}).$$

The vector $\mathbf{P} = \{P_i\}$ which was introduced in Section 1.5 has the following components at M:

$$\mathbf{P} = \left\{ \begin{vmatrix} \xi_{1x_2} & \xi_{1x_3} \\ \xi_{2x_2} & \xi_{2x_3} \end{vmatrix}, \begin{vmatrix} \xi_{1x_3} & \xi_{1x_1} \\ \xi_{2x_3} & \xi_{2x_1} \end{vmatrix}, \begin{vmatrix} \xi_{1x_1} & \xi_{1x_2} \\ \xi_{2x_1} & \xi_{2x_2} \end{vmatrix} \right\}. \tag{6}$$

Thus, the third component of **P**, i.e. the projection of **P** onto $\mathbf{n}(M)$, is equal to K. Transform the first and the second components of **P** as follows. Add and subtract $\xi_{1x_1}\xi_{1x_3}$ to P_1, $\xi_{2x_2}\xi_{2x_3}$ to P_2. We obtain

$$P_1 = \xi_{1x_1}\xi_{1x_3} + \xi_{1x_2}\xi_{2x_3} + 2H\xi_{1x_3},$$
$$P_2 = \xi_{2x_1}\xi_{1x_3} + \xi_{2x_2}\xi_{2x_3} + 2H\xi_{2x_3}.$$

The vector with components $\{\xi_{1x_3}, \xi_{2x_3}, 0\}$ is the vector of curvature of the **n** field streamline (see Section 1.3). Denote it by **k** . Now consider the following vector:

$$\mathbf{l} = \{\xi_{1x_1}\xi_{1x_3} + \xi_{1x_2}\xi_{2x_3}, \xi_{2x_1}\xi_{1x_3} + \xi_{2x_2}\xi_{2x_3}, 0\}.$$

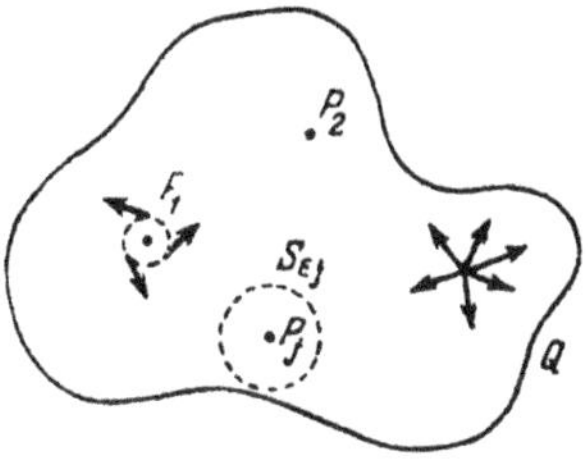

FIGURE 11

If **a** and **b** are vector fields of components a_j and b_j respectively with respect to Cartesian coordinates then $\nabla_{\mathbf{a}}\mathbf{b}$ means, as usual, the vector of components $(\nabla_{\mathbf{a}}\mathbf{b})_i = \frac{\partial b_i}{\partial x_j} a_j$. We call $\nabla_{\mathbf{a}}\mathbf{b}$ the derivative of **b** along **a**.

It is easy to see that **l** is the derivative of **n** along **k**, i.e. $\nabla_{\mathbf{k}}\mathbf{n} = \mathbf{l}$.

Thus, **P** *can be represented as*

$$\mathbf{P} = \mathbf{n}K + 2H\mathbf{k} + \nabla_{\mathbf{k}}\mathbf{n}.$$

Substituting the latter expression into (5) we obtain (3). If there are no singular points of **n** inside F then $\theta = 0$. Indeed, contract the surface F with a continuous deformation to a sufficiently small neighborhood of some intrinsic point x_0 where **n** differs slightly from $\mathbf{n}(x_0)$. The image of F after deformation will be located in some small neighborhood in S^2. Therefore, the degree of ψ is zero. This fact also follows from the equation $\operatorname{div}\mathbf{P} = 0$.

Suppose that there is a finite number of singular points $P_1, \ldots, P_l$ inside Q at which the field **n** loses regularity. Then we can define the index I_j of each singular point P_j. To do this take a sphere S_{ε_j} of radius ε containing only one singular point (see Fig. 11). The *index* I_j of a singular point P_j with respect to the field **n** is the degree of mapping of S_{ε} onto S^2 by means of the field **n**.

Suppose that F contains the field **n** singular points $P_1, \ldots, P_l$. Then F together with S_{ε_j} bounds a domain Q_{ε}. Since this domain does not contain any singular point,

$$\int_F (\mathbf{P}, \nu)\, dS + \sum_{j=1}^{l} \int_{S_{\varepsilon_j}} (\mathbf{P}, \nu)\, dS = 0,$$

where the normal ν on S_{ε_j} is directed inside S_{ε_j}. Therefore, each integral over S_{ε_j} is equal to the degree of mapping S_{ε_j} onto S^2 with the negative sign. Hence, if the surface F encloses the singular points of the vector field **n** then

$$\int_F (\mathbf{P}, \nu)\, dS = \sum_{j=1}^{l} I_j.$$

1.11 The Gauss–Bonnet Formula for the Case of a Surface with a Boundary

Now consider a regular surface F having a regular boundary Γ. The image of this surface in the unit sphere is some domain g with boundary γ, where γ is the image of Γ under the mapping ψ (see Fig. 12). The domain g we shall consider as a many-valued Riemannian surface. The surface F is partitioned into connected domains G_i in which $(\mathbf{P}, \nu)$ preserves the sign. If $(\mathbf{P}, \nu) > 0$ in some domain G_i then we take its spherical image with a $+$ sign . If $(\mathbf{P}, \nu) < 0$ in some domain G_i then we take its spherical image area with a $-$ sign. We call the algebraic sum of areas of all g_i the spherical image area of the surface F. Denote it by $\sigma(g)$.

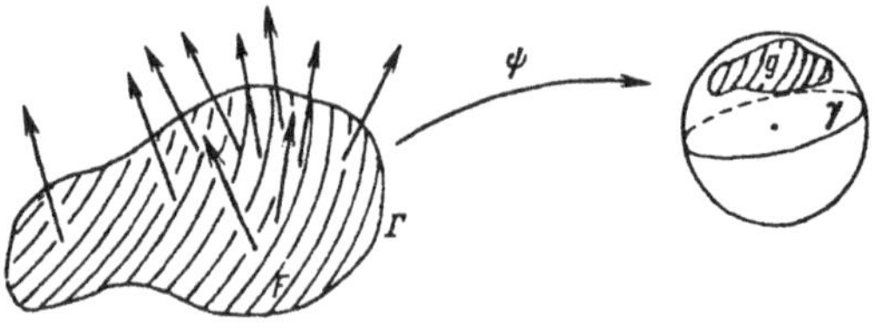

FIGURE 12

Applying the Gauss–Bonnet formula to the γ bounded domain g in the unit sphere S^2, represent the $\sigma(g)$ in terms of the contour integral of γ geodesic curvature

$$\sigma(g) = \int_{\gamma} k_g(\gamma)\, d\tau + 2\pi m, \tag{1}$$

where $k_g(\gamma)$ means the geodesic curvature of γ in S^2, τ means the arc length parameter of γ.

The geodesic curvature of γ can be expressed in terms of the vector field $\mathbf{n}$ if we recall its definition. Let s be the arc length parameter of Γ. Then

$$k_g(\gamma) = -(\mathbf{n}, \mathbf{n}_\tau, \mathbf{n}_{\tau\tau}) = -(\mathbf{n}, \mathbf{n}_s, \mathbf{n}_{ss})\left(\frac{ds}{d\tau}\right)^3.$$

Since $\mathbf{n}_\tau^2 = 1$ and $\mathbf{n}_s = \mathbf{n}_\tau \frac{d\tau}{ds}$, $\left(\frac{ds}{d\tau}\right)^2 = \frac{1}{n_s^2}$. Hence, we can represent $\sigma(g)$ in terms of the integral along γ of some value which depends on $\mathbf{n}$. Applying (4) of Section 1.10, we obtain

$$\int_F (\mathbf{P}, \nu)\, dS = -\int_\Gamma \frac{(\mathbf{n}, \mathbf{n}_s, \mathbf{n}_{ss})}{\mathbf{n}_s^2}\, ds + 2\pi m. \tag{2}$$

We may consider this relation as the analogue of formula (1) of Section 1.10. Now consider the cases when the integrand in (2) can be expressed in terms of the usual geometric values.

Remember that by definition from Section 1.4 the geodesic curvature of Γ with respect to the vector field **n** is the value $\frac{1}{\rho_g} = -(\mathbf{n}, \mathbf{x}_s, \mathbf{x}_{ss})$, where $\mathbf{x} = \mathbf{x}(s)$ is a position vector of Γ with respect to the arc length parameter s.

The following theorem holds.

Theorem *Let Γ be a regular boundary of a surface F. If the field* **n** *is orthogonal to Γ at the points of Γ then*

$$\int_F (\mathbf{P}, \nu)\, dS = \int_\Gamma \frac{ds}{\rho_g} + 2\pi k,$$

where k is an integer.

To prove the theorem we use (2) and the following lemma [25].

Lemma Let $\xi(t)$, $\eta(t)$ be C^2 regular vector functions defined in the segment $0 \le t \le T$ such that $\xi(t)$ and $\eta(t)$ are three-dimensional unit vectors mutually orthogonal for each t. Let φ be the angle between ξ_t and η. Then

$$\int_0^T \frac{(\xi, \xi_t, \xi_{tt})}{\xi_t^2}\, dt = \int_0^T (\xi, \eta, \eta_t)\, dt + \int_0^T \frac{d\varphi}{dt}\, dt.$$

Set $\nu = [\xi, \eta]$. Let us take ξ, η, ν as a basis in E^3. For each t the following holds:

$$\xi_t = \lambda(\cos\varphi\eta + \sin\varphi\nu),$$

$$\xi_{tt} = \lambda_t(\cos\varphi\eta + \sin\varphi\nu) + \lambda(-\sin\varphi\eta + \cos\varphi\nu)\frac{d\varphi}{dt} + \lambda(\cos\varphi\eta_t + \sin\varphi\nu_t).$$

We have

$$\xi_t^2 = \lambda^2,$$

$$(\xi, \xi_t, \xi_{tt}) = \lambda^2\left\{\frac{d\varphi}{dt} + \cos^2\varphi(\xi, \eta, \eta_t) + \sin^2\varphi(\xi, \nu, \nu_t)\right.$$

$$\left. + \cos\varphi\sin\varphi\,((\xi, \eta, \nu_t) + (\xi, \nu, \eta_t))\right\}.$$

From the definition of ν we see $[\xi, \eta] = \nu$, $[\xi, \nu] = -\eta$. Therefore,

$$(\xi, \nu, \nu_t) = (\xi, \eta, \eta_t), \quad (\xi, \eta, \nu_t) = (\nu, \nu_t) = 0, \quad (\xi, \nu, \eta_t) = -(\eta, \eta_t) = 0.$$

Hence

$$\frac{(\xi, \xi_t, \xi_{tt})}{\xi_t^2} = \frac{d\varphi}{dt} + (\xi, \eta, \eta_t).$$

The lemma is proved.

Now consider the curve Γ and the vector field **n**. Suppose that Γ is orthogonal to **n** at each of its points. Apply the lemma, taking ξ as **n** and η as $\mathbf{x}_s$, i.e. as a unit tangent to Γ. By the hypothesis of the theorem, $(\mathbf{n}, \mathbf{x}_s) = 0$. If φ is the angle between $\mathbf{n}_s$ and $\mathbf{x}_s$ then by the lemma we have

$$\int_\Gamma \frac{(\mathbf{n}, \mathbf{n}_s, \mathbf{n}_{ss})}{\mathbf{n}_s^2}\, ds = \int_\Gamma (\mathbf{n}, \mathbf{x}_s, \mathbf{x}_{ss})\, ds + \int_\Gamma \frac{d\varphi}{ds}\, ds = -\int_\Gamma \frac{ds}{\rho_g} + 2\pi n,$$

where n is an integer, $1/\rho_g$ is the geodesic curvature of Γ with respect to $\mathbf{n}$. So, the integral on the right in (2) is expressed in terms of the integral of the geodesic curvature of Γ. The theorem is proved.

The second interesting case occurs when Γ is a closed streamline of $\mathbf{n}$.

Theorem *Let Γ be a closed streamline of the field* $\mathbf{n}$. *If* κ *is a torsion of* Γ *and* F *is a surface with* Γ *as a boundary then*

$$\int_F (\mathbf{P}, \nu)\, dS = -\int_\Gamma \kappa\, ds + 2\pi m. \tag{4}$$

Consider (2). In our case $\mathbf{n} = \mathbf{x}_s$, where $\mathbf{x}_s$ is tangent to Γ. So we have

$$\int_\Gamma \frac{(\mathbf{n}, \mathbf{n}_s, \mathbf{n}_{ss})}{\mathbf{n}_s^2}\, ds = \int_\Gamma \frac{(\mathbf{x}_s, \mathbf{x}_{ss}, \mathbf{x}_{sss})}{\mathbf{x}_{ss}^2}\, ds = \int_\Gamma \kappa\, ds,$$

which settles (4).

Finally, for the third case we suppose that Γ is a closed curve with $\mathbf{n}$ being constant along it.

If $\mathbf{P} \neq 0$ at some point from the domain of definition of $\mathbf{n}$ then one and only one curve tangential to $\mathbf{P}$ with the required property passes through this point. The curves $\mathbf{n} = \text{const}$ are the streamlines of $\mathbf{P}$.

Let F be some surface based on Γ. The image in S^2 of F under the mapping ψ has a boundary consisting of a single point because $\mathbf{n}$ is constant along Γ. Hence, the $\psi(F)$ covers the sphere an integer number times. The following theorem holds.

Theorem *If* Γ *is a closed curve such that* $\mathbf{n} = \mathit{const}$ *along it then for any surface* F *based on* Γ

$$\int_F (\mathbf{P}, \nu)\, dS = 4\pi\theta, \tag{5}$$

where θ *is an integer.*

Let us give the interpretation of θ in the case when the field is defined at all points of E^3 and there is a limit of $\mathbf{n}(M)$ at infinity. In this case we can put the continuous mapping φ of the sphere S^3 onto the sphere S^2 into correspondence with the field $\mathbf{n}$. We consider S^3 as the usual sphere in E^4. At the north pole of it we take a tangent space E^3 where the field $\mathbf{n}$ is defined. We denote the stereographic projection of S^3 onto E^3 from the south pole by α. This projection is formed with the rays from the south pole. We put the point P of the ray intersection with S^3 into correspondence with the point Q of intersection of the same ray with E^3. The south pole of S^3

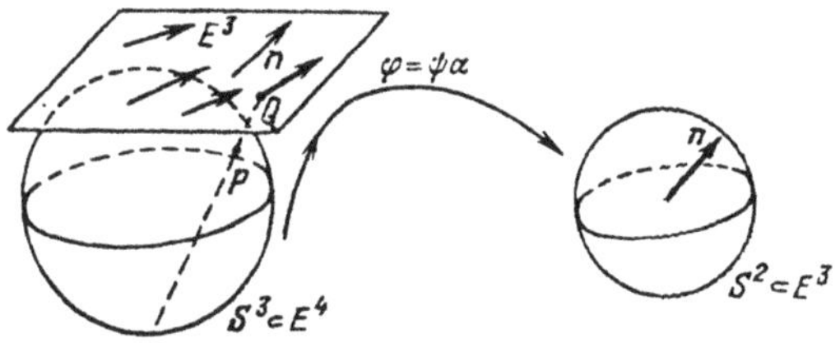

FIGURE 13

corresponds to a point at infinity in E^3. We can put each $Q \in E^3$ into correspondence with a point A in the unit sphere $S^2 \subset E^3$ of the center at some point O. To do this, we put $\mathbf{n}(Q)$ at O and denote the end-point by A. Denote the latter mapping of E^3 onto S^2 by ψ. By hypothesis, the vector field $\mathbf{n}$ is defined at all points of E^3 and also at infinity. The composite mapping $\varphi = \psi\alpha$ defines the continuous mapping of S^3 onto S^2 (see Fig. 13). So, each continuous vector field $\mathbf{n}$, defined in E^3 including infinity, defines a continuous mapping of S^3 onto S^2. The inverse statement is also true: each continuous mapping of S^3 onto S^2 we can put into correspondence with a unit vector field in E^3 having a limit value at infinity. To do this, if $A \in S^2$ corresponds to $P \in S^3$ then define the vector of the field at $Q \in E^3$ setting $\mathbf{n} = \overline{OA}$.

The continuous vector fields $\mathbf{n}_1(x)$ and $\mathbf{n}_2(x)$ defined in $E^3 + \infty$ are said to be *homotopic* if there is a continuous family of vector fields $\mathbf{n}(x, t)$, $0 \leq t \leq 1$ such that $\mathbf{n}(x, 0) = \mathbf{n}_1(x)$, $\mathbf{n}(x, 1) = \mathbf{n}_2(x)$. The vector fields $\mathbf{n}_1(x)$ and $\mathbf{n}_2(x)$ are homotopic if and only if the corresponding mappings ψ_1 and ψ_2 of S^3 onto S^2 are homotopic. Various continuous mappings of S^3 onto S^2 can be separated into classes of homotopic mappings. We know from topology the following theorem (see [78], p. 98).

Theorem *The homotopy classes of mappings φ: $S^3 \to S^2$ are in one-to-one correspondence to integers. For any integer γ the corresponding homotopy class is a class of mappings $\varphi = p\Phi$: $S^2 \to S^3$ where p: $S^3 \to S^2$ is the Hopf mapping and Φ: $S^3 \to S^3$ is a mapping of degree γ.*

The number γ is called the *Hopf invariant* of φ: $S^3 \to S^2$. For any point $P \in S^3$ a linear mapping ∇_φ of tangent space $T_p(S^3)$ into the tangent space $T_A(S^2)$ is defined. We say that φ is *normal at P* if the image of the mapping above coincides with $T_A S^2$. We say that φ is *normal over* $A \in S^2$ if φ is normal at any point $P \in \varphi^{-1}(A)$. From topology (see [51], p. 206) we know that for any point $A \in S^2$ there is however close to φ a smooth mapping $\tilde{\varphi}$ which is homotopic to φ and normal over A. Further on we suppose that φ is normal over A. It can be proved that if φ is normal over A then the inverse image $\Gamma_A = \varphi^{-1}(A)$ is a set of smooth curves in S^3. To find the Hopf invariant we can apply the Whitehead formula. Orient the curve in Γ_A as follows. Choose in the tangent space of S^3 at P a basis $\mathbf{e}_1, \mathbf{e}_2, \mathbf{e}_3$ which defines a positive orientation of S^3 and such that $\nabla_\varphi \mathbf{e}_1 = 0$ while $\nabla_\varphi \mathbf{e}_2$, $\nabla_\varphi \mathbf{e}_3$ defines a positive orientation in S^2; then $\mathbf{e}_1$ will define a positive orientation of Γ_A. Let F be a smooth surface in S^3 based on Γ_A such that its orientation is coordinated with the orientation of Γ_A. Let ω be an area form of S^2 and $\varphi^*\omega$ be an induced form on S^3. Then for the Hopf invariant γ we have

$$\gamma = \frac{1}{4\pi}\int_F \varphi^*\omega = \frac{1}{4\pi}\int_{\varphi F} \omega.$$

Since φ maps the boundary of F, i.e. the Γ_A, into a single point, the image of F under the mapping φ is a two-dimensional cycle on S^2. The degree of mapping φ restricted to F is defined and is equal to the Hopf invariant.

Turning back to (5), we note that θ will be the Hopf invariant if the boundary of F is a pre-image of some point $A \in S^2$, i.e. can be constituted from some number of closed curves with an orientation described as above.

Consider now in three-dimensional Euclidean space E^3, with a point at infinity included, the unit vector field **n** generated by the Hopf mapping. The Hopf mapping is constructed as follows. Represent the three-dimensional sphere S^3 as a unit sphere in complex space C^2 of two complex variables z_1, z_2, i.e. as a set of points $(z_1, z_2) \in C^2$ such that

$$z_1\bar{z}_1 + z_2\bar{z}_2 = 1,$$

where $\bar{z}_i$ means the complex conjugate to z_i. At the same time, the two-dimensional sphere S^2 can be represented as a complex projective straight line, i.e. as a set of pairs (z_1, z_2) not being zero simultaneously complex numbers and such that (z_1, z_2) and (z'_1, z'_2) are identical to each other if there is a complex $\lambda \neq 0$ such that $z_1 = \lambda z'_1$, $z_2 = \lambda z'_2$. We denote the class of identical pairs (z_1, z_2) with respect to the equivalence above by $[z_1, z_2]$. The Hopf mapping p: $S^3 \to S^2$ is defined by the following formula:

$$p(z_1, z_2) = [z_1, z_2],$$

where $(z_1, z_2) \in S^2$. In this mapping the inverse image of each point of S^2 is a great circle S^1 in S^3. The three-dimensional sphere S^3 is decomposed in a family of great circles; moreover, as the quotient space of that decomposition we have the two-dimensional sphere S^2. Now construct the *vector field in E^3 which is generated by the Hopf mapping*. Since we can put any point of S^3 into correspondence with a point in E^3, excluding a point at infinity, and we can put a point of S^2 into correspondence with a vector of constant length, the mapping p: $S^3 \to S^2$ generates some vector field in E^3. At infinity we are also able to obtain a definite vector. Proceed to the vector field construction. Let S^3 be a sphere of unit radius in E^4 defined by the equation $x_1^2 + \cdots + x_4^2 = 1$, where x_i are Cartesian coordinates in E^4. Let E^3 be the tangent space to the sphere at a point $(0, 0, 0, 1)$ and y_i be Cartesian coordinates in that E^3. By means of a stereographic projection from $(0, 0, 0, -1)$ we put the point in E^3 into correspondence with a point (x_1, x_2, x_3, x_4). The coordinates of the corresponding point in E^3 are as follows:

$$y_i = \frac{2x_i}{x_4 + 1}, \quad i = 1, 2, 3.$$

Set $z_1 = x_1 + ix_2$, $z_2 = x_3 + ix_4$. Suppose that $z_2 \neq 0$. Next we put the point in S^2 into correspondence with the point $(x_1, \dots, x_4) \in S^3 : \left[\frac{z_1}{z_2}, 1\right]$. The complex number $\frac{x_1+ix_2}{x_3+ix_4} = u + iv$, provided that $x_3 + ix_4 \neq 0$, has the corresponding point in some plane E^2 tangent to S^2. We put the point $u + iv$ in E^2 into correspondence with the vector ξ having components

$$\xi_1 = \frac{2u}{1 + u^2 + v^2}, \quad \xi_2 = \frac{2v}{1 + u^2 + v^2}, \quad \xi_3 = \frac{u^2 + v^2 - 1}{1 + u^2 + v^2}.$$

If we put ξ into the origin then the ξ end-point defines the point in S^2. Represent the components of ξ as a functions of y_i. At first, represent ξ_i in terms of x_i. We have

$$u + iv = \frac{x_1x_3 + x_2x_4 + i(x_2x_3 - x_1x_4)}{x_3^2 + x_4^2}.$$

From this it follows that $u^2 + v^2 = \frac{x_1^2+x_2^2}{x_3^2+x_4^2}$. Therefore,

$$\xi_1 = \frac{2u}{1 + u^2 + v^2} = \frac{2(x_1x_3 + x_2x_4)}{x_1^2 + x_2^2 + x_3^2 + x_4^2}.$$

Taking into account that $\sum_{i=1}^{4} x_i^2 = 1$, we obtain

$$\xi_1 = 2(x_1x_3 + x_2x_4).$$

In the same way we find

$$\xi_2 = 2(x_2x_3 - x_1x_4).$$

Substituting the expression for $u^2 + v^2$ above into the expression for ξ_3, we have

$$\xi_3 = x_1^2 + x_2^2 - x_3^2 - x_4^2.$$

Now express x_i in terms of y_i. It is evident that

$$1 - x_4^2 = \sum_{i=1}^{3} x_i^2.$$

Set

$$a = \sum_{i=1}^{3} y_i^2 = \frac{4\sum_{i=1}^{3} x_i^2}{(1 + x_4)^2} = \frac{4(1 - x_4)}{(1 + x_4)}.$$

From this we find the expressions of x_4 and $x_4 + 1$ in terms of a:

$$x_4 = \frac{4 - a}{4 + a}, \quad x_4 + 1 = \frac{8}{4 + a}.$$

Then we have

$$x_i = \frac{y_i(x_4 + 1)}{2} = \frac{4y_i}{4 + a}, \quad i = 1, 2, 3.$$

Substituting the expressions above into the ones for ξ_i, we obtain

$$\xi_1 = \frac{8(4y_1y_3 + y_2(4-a))}{(4+a)^2}, \quad \xi_2 = \frac{8(4y_2y_3 - y_1(4-a))}{(4+a)^2},$$

$$\xi_3 = \frac{16(y_1^2 + y_2^2 - y_3^2) - (4-a)^2}{(4+a)^2}.$$

Consider the behavior of this vector field. If $y_1^2 + y_2^2 + y_3^2 \to \infty$ then $\xi_1 \to 0$, $\xi_2 \to 0$, $\xi_3 \to -1$. Thus, when the point tends to infinity, the vector field ξ tends to a definite limit. Along the y_3-axis the vector field ξ is constant, namely $(0, 0, -1)$. It is easy to see that along the circle $y_1^2 + y_2^2 = \text{const}$ the component ξ_3 stays constant. From the expressions for ξ_1 and ξ_2 we see that ξ goes into itself in rotating around the y_3-axis, i.e. ξ is invariant with respect to rotation around the y_3-axis. Therefore, if we know the field behavior in the y_1, y_2-plane then we are able to reconstruct the field behavior in space. Consider the field behavior along the y_1-axis. If $y_2 = y_3 = 0$ then

$$\xi_1 = 0, \quad \xi_2 = \frac{-8y_1(4 - y_1^2)}{(4 + y_1^2)^2}, \quad \xi_3 = \frac{16y_1^2 - (4 - y_1^2)^2}{(4+a)^2}.$$

Thus, along the y_1-axis the vectors of the field ξ lie in the plane which is orthogonal to this axis. In moving along this axis ξ is rotating in that plane clockwise (from the $+\infty$ viewpoint). If $0 < y_1 < 2$ then $\xi_2 < 0$ and if $y_1 > 2$ then $\xi_2 > 0$. Therefore, in varying y_1 from 0 to $+\infty$ the ξ turns around in an angle 2π. At the points P_1 and P_2 in a half-axis $[0, \infty)$ with $y_1^2 = 4(3 - 2\sqrt{2})$ and $y_1^2 = 4(3 + 2\sqrt{2})$ respectively we have $\xi_3 = 0$. Also, at P_1 the field vector ξ is directed opposite to the direction of the y_2-axis, while at P_2 it is directed along the y_2-axis. The circles in the y_1y_2-plane centered at the origin which are the vector field ξ streamlines pass through these points. However, they have the opposite bypass to each other in moving along the ξ direction.

The total rotation of the vector field ξ vectors in varying y_1 from $-\infty$ to $+\infty$ is equal to 4π (see Fig. 14).

Consider the streamlines of that vector field. They satisfy the following system of differential equations:

$$\frac{dy_i}{ds} = \xi_i, \quad i = 1, 2, 3.$$

Suppose that a point moves along the ξ streamline in the ξ direction. From the first two equations we find

$$\frac{d \arctan \frac{y_2}{y_1}}{ds} = \frac{a-4}{(a+4)^2}, \quad \frac{d(y_1^2 + y_2^2)}{ds} = \frac{64(y_1^2 + y_2^2)y_3}{(4+a)^2}.$$

From this it follows that if a point on the streamline lies inside the ball $y_1^2 + y_2^2 + y_3^2 < 4$ then its projection into the y_1y_2-plane moves clockwise; if a point lies outside that ball then its projection moves in the opposite direction. When $y_3 \to \infty$, the streamlines turn around the y_3-axis infinitely many times. When $y_3 > 0$

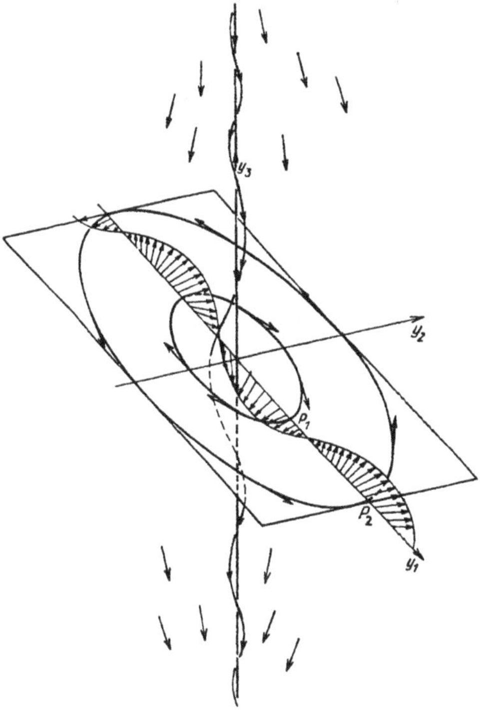

FIGURE 14

the distance between the projection and the y_3-axis increases, while for $y_3 < 0$ that distance decreases. Consider the behavior of y_3 along the streamline. We have $dy_3/ds = 0$ if $y_3^2 = -(4+\rho) + 4\sqrt{2\rho}$. The latter equation represents the convex curve with the end-points in the ρ-axis in a plane of parameters y_3^2, ρ. This curve corresponds to some closed curve γ in the y_3, y_1-plane which generates a torus T in rotating around the y_3-axis. A streamline emitted from a point in the annulus, bounded by the circles $\rho = 4(3-2\sqrt{2})$ and $\rho = 4(3+2\sqrt{2})$ in the plane $y_3 = 0$, rises at first and then descends.

1.12 The Extremal Values of Geodesic Torsion

The torsion of a geodesic (straightest) line having a given direction is what we call the *geodesic torsion of the field in that direction*. Let dr/ds be the tangent vector of a geodesic line. This principal normal of this curve coincides with the field **n**. Denote the binormal by **b**. Then by the Frenet formulas we get

$$\frac{d\mathbf{n}}{ds} = -k\frac{d\mathbf{r}}{ds} + \kappa\mathbf{b},$$

where k is the curvature and κ is the torsion of the straightest line. From this we get *the expression for the geodesic torsion*:

$$\kappa = \frac{(d\mathbf{r}, \mathbf{n}, d\mathbf{n})}{ds^2}. \tag{1}$$

Thus, *the geodesic torsion with respect to the principal direction of the second kind is zero*. Let κ_1 and κ_2 be the extremal values of (1) in rotating dr/ds in a plane which is orthogonal to **n**. Then, as Rogers stated, $\kappa_1 + \kappa_2 = (\mathbf{n}, \operatorname{curl}\mathbf{n})$. Prove this. Choose the basis in such a way that $\xi_1 = \xi_2 = 0$, $\xi_3 = 1$. The expression (1) obtains the following form

$$\kappa = \frac{1}{ds^2}\begin{vmatrix} d\xi_1 & d\xi_2 & d\xi_3 \\ dx_1 & dx_2 & 0 \\ 0 & 0 & 1 \end{vmatrix} = \frac{d\xi_1}{ds}\frac{dx_2}{ds} - \frac{d\xi_2}{ds}\frac{dx_1}{ds}$$

$$= -\xi_{2x_1}\left(\frac{dx_1}{ds}\right)^2 - (\xi_{2x_2} - \xi_{1x_1})\frac{dx_2}{ds}\frac{dx_1}{ds} + \xi_{1x_2}\left(\frac{dx_2}{ds}\right)^2.$$

To find the extremal values of κ, form the symmetric matrix and the characteristic equation in the following manner:

$$\begin{vmatrix} \xi_{2x_1} - \lambda & \frac{\xi_{2x_2} - \xi_{1x_1}}{2} \\ \frac{\xi_{2x_2} - \xi_{1x_1}}{2} & -\xi_{1x_2} - \lambda \end{vmatrix} = 0.$$

In expanded form we have

$$\lambda^2 - \lambda(\xi_{2x_1} - \xi_{1x_2}) - \left[\frac{(\xi_{2x_2} - \xi_{1x_1})^2}{4} + \xi_{1x_2}\xi_{2x_1}\right] = 0.$$

From this we find that the extremal values of the torsion satisfy the equation

$$\lambda^2 - 2\lambda\rho - (H^2 - K) = 0,$$

$$\kappa_{1,2} = \rho \pm \sqrt{\rho^2 + H^2 - K}.$$

So, we get

$$\kappa_1 + \kappa_2 = 2\rho = (\mathbf{n}, \operatorname{curl}\mathbf{n}).$$

It is possible to consider (1) with respect to arbitrary directions of dr/ds, i.e. not only those which are orthogonal to **n**. Let τ_i, $i = 1, 2, 3$ be the extremal values of (1). We call them the *principal geodesic torsions*. Introduce the *total*, *mixed* and *mean geodesic torsions* as

$$T = \tau_1\tau_2\tau_3, \quad M = \tau_1\tau_2 + \tau_1\tau_3 + \tau_2\tau_3, \quad S = \tau_1 + \tau_2 + \tau_3.$$

respectively. It happens that $S = (\mathbf{n}, \operatorname{curl} \mathbf{n})$. As above, we consider (1) with respect to the special choice of basis:

$$\kappa = \begin{vmatrix} d\xi_1 & d\xi_2 & d\xi_3 \\ dx_1 & dx_2 & dx_3 \\ 0 & 0 & 1 \end{vmatrix} = \frac{d\xi_1\, dx_2 - d\xi_2\, dx_1}{ds^2}$$

$$= -\xi_{2x_1}\left(\frac{dx_1}{ds}\right)^2 - \xi_{2x_2}\frac{dx_2}{ds}\frac{dx_1}{ds} - \xi_{2x_3}\frac{dx_3}{ds}\frac{dx_1}{ds}$$

$$+\xi_{1x_1}\frac{dx_1}{ds}\frac{dx_2}{ds} + \xi_{1x_2}\left(\frac{dx_2}{ds}\right)^2 + \xi_{1x_3}\frac{dx_3}{ds}\frac{dx_2}{ds}.$$

We can find the extremal values from the following characteristic equation:

$$\begin{vmatrix} -\xi_{2x_1} - \lambda & \frac{\xi_{1x_1} - \xi_{2x_2}}{2} & \frac{-\xi_{2x_3}}{2} \\ \frac{\xi_{1x_1} - \xi_{2x_2}}{2} & \xi_{1x_2} - \lambda & \frac{\xi_{1x_3}}{2} \\ \frac{-\xi_{2x_3}}{2} & \frac{\xi_{1x_3}}{2} & -\lambda \end{vmatrix} = 0.$$

In expanded form we have

$$-\lambda^3 + \lambda^2(\xi_{1x_2} - \xi_{2x_1}) + \lambda\left\{\xi_{2x_1}\xi_{1x_2} + \frac{(\xi_{1x_1} - \xi_{2x_2})^2 + \xi_{1x_3}^2 + \xi_{2x_3}^2}{4}\right\}$$

$$+\begin{vmatrix} -\xi_{2x_1} & \frac{\xi_{1x_1} - \xi_{2x_2}}{2} & \frac{-\xi_{2x_3}}{2} \\ \frac{\xi_{1x_1} - \xi_{2x_2}}{2} & \xi_{1x_2} & \frac{\xi_{1x_3}}{2} \\ \frac{-\xi_{2x_3}}{2} & \frac{\xi_{1x_3}}{2} & 0 \end{vmatrix} = 0.$$

Thus, *for the mean and the mixed torsions of the field we have*

$$S = (\mathbf{n}, \operatorname{curl} \mathbf{n}), \quad M = K - H^2 - \frac{|[\mathbf{n}, \operatorname{curl} \mathbf{n}]|}{4}.$$

The value T represents some field invariant. With respect to the chosen coordinates we have

$$T = \frac{1}{4}\left\{-\xi_{2x_3}^2\xi_{1x_2} + (\xi_{2x_2} - \xi_{1x_1})\xi_{1x_3}\xi_{2x_3} + \xi_{1x_3}^2\xi_{2x_1}\right\}. \tag{2}$$

1.13 The Singularities as the Sources of Curvature of a Vector Field

Suppose that there are singular points of the field **n** in a domain G. We denote the set of singular points by M. Introduce a geometrical characteristic of the singular points. Let T_ε be the ε-neighborhood of M and F_ε be the boundary of T_ε. Formula (2) of Section 1.8, which gives the expression for the integrated field curvature K over the volume, has been stated on the supposition that there are no singular points inside the surface F. Let ν be the normal of F_ε directed inward to T_ε. We say that *the singularity is the curvature source of power Q if there is the following limit*:

$$Q = \lim_{\varepsilon \to 0} \frac{1}{2} \int_{F} \{(\mathbf{n}, \nu) \operatorname{div} \mathbf{n} - (\mathbf{k}, \nu)\} \, dS. \tag{1}$$

If there is the curvature source in G with F as the boundary then we shall consider the integral of K over G as an improper integral, i.e. set

$$\int_G K \, dV = \lim_{\varepsilon \to 0} \int_{G \setminus T} K \, dV.$$

It is possible to rewrite formula (2) in Section 1.8 with regard for singular points inside G as

$$\int_G K \, dV = -\int_{\partial G} \{(\mathbf{n}, \nu) H + (\mathbf{k}, \nu)\} \, dS + Q.$$

Thus, if the vector field **n** is fixed on F then the higher the field source power the more the integrated curvature K is enclosed inside F.

There may be an arbitrarily powerful source enclosed in an arbitrarily small volume. For instance, take some closed curve γ of length l in a domain G. Define the vector field **n** in a neighborhood of γ as a tangent vector field for the family of concentrated circles in the planes normal to γ with their centers in γ. As we have stated in Section 1.8, the integral on the right of (2) is equal to $-\pi l$, i.e. the curve is the curvature source of power $-\pi l$. Note that ν is directed inside the tube.

Consider the isolated singular point M_0. *If the singularity is such that for any ε however small the modulus of the area of the image of the sphere F_ε of radius ε centered at M_0 is bounded, i.e.*

$$\int_{\psi(F_\varepsilon)} |d\sigma| < C = \text{const},$$

then the curvature source power is zero.

Indeed, use the relation, stated in Section 1.8, between the integral on the right of (1) and the integral of $(\mathbf{x}, \mathbf{n})$ over $\psi(F_\varepsilon)$. It is also possible to write

$$Q = \lim_{\varepsilon \to 0} \int_{\psi(F_\varepsilon)} (\mathbf{x}, \mathbf{n}) \, d\sigma.$$

Put the origin into M_0. Since $|(\mathbf{x}, \mathbf{n})| < \varepsilon$ over ∂F_ε, then

$$\left| \int\limits_{\partial(F_\varepsilon)} (\mathbf{x}, \mathbf{n})\, d\sigma \right| \le \varepsilon C.$$

Passing to the limit when $\varepsilon \to 0$, we obtain $Q = 0$. From this it follows, for instance, that *the isolated singular points of algebraic fields have zero power.*

Let the vector field $\mathbf{n}$ be singular at the points of some closed curve γ; moreover, we suppose that at each point $P \in \gamma$ the field $\mathbf{n}$ is in the normal plane of γ and is directed along the radius of the circle centered at P in that plane.

Suppose that the field $\mathbf{n}$ is directed outward at the points of a tubular surface F of sufficiently small radius R with γ as an axial curve. In this case $(\mathbf{n}, \nu) = -1$. Let us find the curvature source power of such a singularity. It is equal to the integrated mean curvature of F with respect to $\mathbf{n}$ directed outward.

Thus, we need to find the mean curvature of the tubular surface with respect to an extrinsic normal. We know from differential geometry that

$$H = \frac{EN - 2MF + GL}{2(EG - F^2)}.$$

Here E, F, G are the coefficients of the first fundamental form and L, M, N are the coefficients of the second fundamental form of the tubular surface. Use the formulas from Section 1.8. We have

$$EG - F^2 = R^2(1 - kR\cos v)^2,$$

$$EN - 2MF + GL = R(1 - kR\cos v)(2kR\cos v - 1).$$

Hence, the mean curvature is

$$H = \frac{2kR\cos v - 1}{2R(1 - kR\cos v)}.$$

So, the curvature source power is

$$Q = \int\limits_F H\, dS = \int\limits_0^{2\pi}\int\limits_0^{l} \frac{2kR\cos v - 1}{2R(1 - kR\cos v)} \sqrt{EG - F^2}\, du\, dv$$

$$= \frac{1}{2}\int\limits_0^{2\pi}\int\limits_0^{l} (2kR\cos v - 1)\, du\, dv = -\pi l,$$

where l is the length of the axial curvature γ. If $\mathbf{n}$ is directed inside the tube then Q gets the opposite sign.

Consider the following problem: in which case the vector field singularity is topologically removable? Let M be a connected set of the singular points of the field having the ε-neighborhood where the vector field $\mathbf{n}$ is defined. Set M_ε to be the ε-neighborhood. We say that *a singularity is topologically removable* if for any sufficiently small $\varepsilon > 0$ the field $\mathbf{n}$ can be smoothly prolongated from ∂M_ε inside M_ε.

Theorem *Suppose that in some ε-neighborhood of the set M of singular points of the field $\mathbf{n}$ the field invariants are bounded as follows:*

$$|K| \leq C_0^2, \quad |H| \leq C_0, \quad |k| \leq C_0, \quad |4(\mathbf{n}, \operatorname{curl} \mathbf{n})T| \leq C_0^4,$$

where T is the total geodesic torsion, $C_0 =$ const. Suppose that the surface measure S of the set M is sufficiently small, namely $S < \frac{4\pi}{\sqrt{13C_0^2}}$. Then the singularity is topologically removable.

Observe that in the case of a holonomic vector field it is sufficient to require the boundness of $|K|$, $|H|$ and $|k|$. To prove the theorem we need to estimate the length of vector $\mathbf{P} = \mathbf{n}K + 2H\mathbf{k} + \nabla_{\mathbf{k}}\mathbf{n}$. Set $\mathbf{l} = \nabla_{\mathbf{k}}\mathbf{n}$. Consider the scalar product of $\mathbf{l}$ and $\mathbf{P}$ in a special choice of coordinate system:

$$(\mathbf{l}, \mathbf{P}) = \begin{vmatrix} \xi_{1x_2} & \xi_{1x_3} \\ \xi_{2x_2} & \xi_{2x_3} \end{vmatrix} (\xi_{2x_3}\xi_{1x_2} + \xi_{1x_1}\xi_{1x_3}) + \begin{vmatrix} \xi_{1x_3} & \xi_{1x_1} \\ \xi_{2x_3} & \xi_{2x_1} \end{vmatrix} (\xi_{1x_3}\xi_{2x_1} + \xi_{2x_2}\xi_{2x_3})$$

$$= (\xi_{1x_2} - \xi_{2x_1})\left[\xi_{2x_3}^2\xi_{1x_2} + (\xi_{1x_1} - \xi_{2x_2})\xi_{1x_3}\xi_{2x_3} - \xi_{1x_3}^2\xi_{2x_1}\right]$$

$$- (\xi_{1x_3}^2 + \xi_{2x_3}^2)(\xi_{1x_1}\xi_{2x_2} - \xi_{1x_2}\xi_{2x_1}).$$

Now use the expression for the total geodesic torsion from Section 1.12. We get

$$(\mathbf{l}, \mathbf{P}) = -4(\mathbf{n}, \operatorname{curl} \mathbf{n})T - K|\mathbf{k}|^2.$$

On the other hand, from the expression for $\mathbf{P}$ above we see that

$$(\mathbf{l}, \mathbf{P}) = +2H(\mathbf{k}, \mathbf{l}) + \mathbf{l}^2.$$

Hence, to find $|\mathbf{l}|$ we have the equality

$$\mathbf{l}^2 + 2H(\mathbf{k}, \mathbf{l}) + 4(\mathbf{n}, \operatorname{curl} \mathbf{n})T + K|\mathbf{k}|^2 = 0.$$

Let φ be the angle between $\mathbf{k}$ and $\mathbf{l}$. Then

$$|\mathbf{l}| = H|\mathbf{k}| \cos\varphi \pm \sqrt{H^2\mathbf{k}^2 \cos^2\varphi - 4(\mathbf{n}, \operatorname{curl} \mathbf{n}) - K|\mathbf{k}|^2}.$$

From this we find that $|\mathbf{l}| \leq 3C_0^2$. Next we consider the square of vector $\mathbf{P}$ length:

$$\mathbf{P}^2 = K^2 + \mathbf{l}^2 + 2H(\mathbf{k}, \mathbf{l}) + 4H\mathbf{k}^2 + 2H(\mathbf{k}, \mathbf{l}).$$

So, the modulus of $\mathbf{P}$ satisfies

$$|\mathbf{P}| \leq 13C_0^2.$$

Apply formula (5) of Section 1.10 to the closed surface ∂M_ε. The area of this surface differs from S arbitrarily small. Since $S < \frac{4\pi}{\sqrt{13C_0^2}}$,

$$\left| \frac{1}{4\pi} \int_{\partial M_\varepsilon} (\mathbf{P}, \nu)\, dS \right| < 1.$$

Therefore, the degree of mapping of ∂M_ε onto the unit sphere S^2 by means of the field $\mathbf{n}$ is equal to zero. As we know (see, for instance, [52], p. 125), in this case the vector field $\mathbf{n}$ can be prolongated without singular points inside ∂M_ε.

1.14 The Mutual Restriction of the Fundamental Invariants of a Vector Field and the Size of Domain of Definition

Let us ask the following question: for a given fixed domain of definition, *is it possible to find a vector field of an arbitrarily large curvature?* The following example gives the answer. Let us consider the cube: $1 \le x_i \le 2$, $i = 1,2,3$. Define the vector field as $\mu\mathbf{n} = (A_1, A_2, A_3)$, where $A_1 = x_1 \cos \lambda x_2$, $A_2 = x_1 \sin \lambda x_2$, $A_3 \equiv 1$, $\lambda > 0$ is some constant. Then by (2) of Section 1.5 we have

$$K = \frac{A_{1x_1} A_{2x_2} - A_{2x_1} A_{1x_1}}{(1 + A_1^2 + A_2^2)^2} = \frac{x_1 \lambda}{(1 + x_1^2)^2} \ge \frac{\lambda}{25}.$$

Thus, we are able to construct in the fixed domain the vector field of arbitrarily large curvature. Observe that in this example the value of non-holonomicity increases together with K increasing. The following theorem holds.

Theorem *If a regular vector field* $\mathbf{n}$ *of curvature* $K \ge K_0 > 0$ *is defined in a ball* D *of radius* r *and the value of the non-holonomicity satisfies* $|(\mathbf{n}, \operatorname{curl} \mathbf{n})| < \rho_0$ *then*

$$r \le \frac{3}{2\sqrt{K_0 - \rho_0^2}}.$$

To prove the theorem we use the inequality stated in Section 1.5:

$$H^2 + \frac{(\mathbf{n}, \operatorname{curl} \mathbf{n})^2}{4} \ge K,$$

which implies that $\operatorname{div} \mathbf{n} \ge 2\sqrt{K_0 - \rho_0^2}$ in the ball D. Integrating the latter inequality inside the ball D and applying the Gauss–Ostrogradski formula, we find

$$\frac{2}{3}\sqrt{K_0 - \rho_0^2}\, \pi r^3 \le \int_D \operatorname{div} \mathbf{n}\, dV = \int_{\partial D} (\nu, \mathbf{n})\, dS \le \pi r^2,$$

where ν is the outward normal of the boundary ∂D of the ball D. Therefore $r \le \frac{3}{2}(K_0 - \rho_0^2)^{-1/2}$.

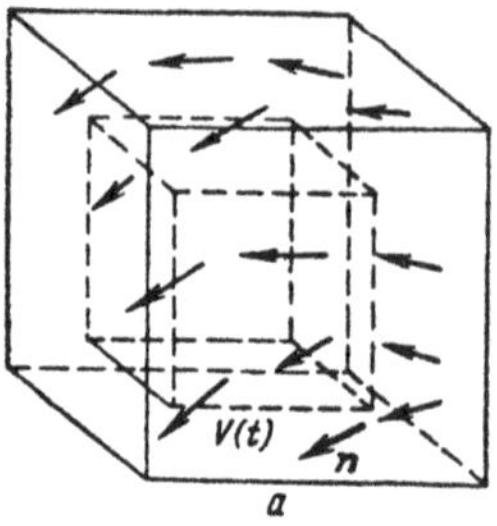

FIGURE 15

Now consider the vector field of large negative total curvature $K < 0$. We shall obtain in this case the restriction on the size of domain of definition provided that the curvature of vector field streamline is bounded from above. The following theorem holds.

Theorem *Let a regular vector field* $\mathbf{n}$ *is defined in a cube of the edge length* a *in Euclidean space* E^3 *such that the total field curvature satisfies the inequality* $K \le -K_0 < 0$, *where* $K_0 = \text{const}$, *and the field streamline curvature is bounded from above* $|k| \le \mu_0$. *Then the cube size is limited, namely*

$$a \le \frac{4\mu_0}{K_0} + \sqrt{\left(\frac{4\mu_0}{K_0}\right)^2 + \frac{24}{K_0}}. \tag{1}$$

To prove the theorem we shall state some integral formula which relates the total curvature and the streamline curvature of the field $\mathbf{n}$. Let $V(t)$ be a cube with the center at the origin and with the edge $2t$ $\{x_i = \pm t\}$ (see Fig. 15). Represent the total curvature K in the divergent form

$$K = \frac{1}{2}\sum_{i,j=1}^{3}\begin{vmatrix} \xi_{ix_i} & \xi_{ix_j} \\ \xi_{jx_i} & \xi_{jx_j} \end{vmatrix} = \frac{1}{2}\left\{\frac{\partial}{\partial x_1}(\xi_1\xi_{2x_2} - \xi_2\xi_{1x_2} + \xi_1\xi_{3x_3} - \xi_3\xi_{1x_3}) + \dots\right\}.$$

By the Gauss–Ostrogradski theorem, it is possible to reduce the integral of K in the cube $V(t)$ to the integral over the surface of the cube. Consider separately the expression which corresponds to the cube face $x_1 = t$. It has the form

$$\frac{1}{2}\iint\limits_{x_1=t}(\xi_1\xi_{2x_2} - \xi_2\xi_{1x_2} + \xi_1\xi_{3x_3} - \xi_3\xi_{1x_3})\,dx_2\,dx_3$$

$$= \frac{1}{2}\iint\limits_{x_1=t}\left(\frac{\partial}{\partial x_2}(\xi_1\xi_2) + \frac{\partial}{\partial x_3}(\xi_1\xi_3) - 2\xi_{1x_2}\xi_2 - 2\xi_{1x_3}\xi_3\right)dx_2\,dx_3. \tag{2}$$

We turn the integrals of the first two terms in this expression into integrals along the edges. To transform the third and the fourth terms we note that we can represent the third and the fourth components of curl **n** as

$$\xi_{3x_1} - \xi_{1x_3} = 2\xi_2\rho + \xi_1 k_3 - \xi_3 k_1,$$

$$\xi_{1x_2} - \xi_{2x_1} = 2\xi_3\rho + \xi_2 k_1 - \xi_1 k_2,$$

where $\rho = \frac{1}{2}(\mathbf{n}, \operatorname{curl}\mathbf{n})$, $\mathbf{k} = (k_1, k_2, k_3)$ is the streamline curvature vector. From this we substitute ξ_{1x_3} and ξ_{1x_2} into (2). Then (2) turns into the sum of integrals along the cube edges and faces:

$$\frac{1}{2}\left(\int_{l_1} \xi_1\xi_2\, dx_3 - \int_{l_2} \xi_1\xi_2\, dx_3 + \int_{l_3} \xi_1\xi_3\, dx_2 - \int_{l_4} \xi_1\xi_3\, dx_2\right)$$

$$-\iint_{x_1=t} \{\xi_2(\xi_{2x_1} + 2\xi_3\rho + \xi_2 k_1 - \xi_1 k_2) + \xi_3(\xi_{3x_1} - 2\xi_2\rho + \xi_3 k_1 - \xi_1 k_3)\}\, dx_2\, dx_3,$$

where l_i are the edges of the face $x_i = t$; for instance, $l_1 : x_1 = t, x_2 = t$. We add and subtract $\xi_1^2 k_1$ to the expression integrated in the face $x_1 = t$. Since $\mathbf{k} \perp \mathbf{n}$, the integrand in the face $x_1 = t$ turns into

$$\frac{1}{2}\frac{\partial}{\partial x_1}(\xi_2^2 + \xi_3^2) + k_1.$$

Take the derivative with respect to x_1 in the first term out of the integral and replace it with the derivative with respect to t. Since the face depends on t, we ought to add the integrals of $(\xi_2^2 + \xi_3^2)/2$ along the edges which bound the face $x_1 = t$. Presenting not the integrals along the edges l_2, l_3 and l_4, we obtain the following expression:

$$\frac{1}{2}\left(\int_{l_1} (\xi_1\xi_2 + \xi_2^2 + \xi_3^2)\, dx_3 + \cdots - \iint_{x_1=t} k_1 dx_2\, dx_3\right) - \frac{1}{2}\frac{d}{dt}\iint_{x_1=t} (\xi_2^2 + \xi_3^2)\, dx_2\, dx_3. \quad (3)$$

Transforming the integrals in the other faces in the same manner, we find that along the edge l_1, for instance, we integrate the following expression:

$$\frac{1}{2}(2\xi_1\xi_2 + \xi_1^2 + \xi_2^2 + 2\xi_3^2) = \frac{1}{2}\left\{(\xi_1 + \xi_2)^2 + 2\xi_3^2\right\} = |a_3|^2.$$

where we denoted by a_i, $i = 1, \ldots, 6$ the projection of **n** into the plane containing the edge l_i and the center of the cube. Let ν be a normal vector of the cube $V(t)$, dS a cube surface area element. So, we have

$$\int_{V(t)} K\, dV = -\frac{d}{dt}\sum_{k=1}^{3}\iint_{\substack{x_k=t\\ k\neq i\neq j}} \frac{\xi_i^2 + \xi_j^2}{2}\, dx_i\, dx_j - \iint_{\partial V} (\mathbf{n}, \operatorname{curl}\mathbf{n}, \nu)\, dS + \sum_{i=1}^{6}\int_{l_i} |a_i|^2\, dx_i. \quad (4)$$

Integrating (4) in t from 0 to $t = a/2$ and taking into account the inequalities

$$\left| \int\limits_{\partial V} (\mathbf{n}, \operatorname{curl} \mathbf{n}, \nu)\, dS \right| \leq \mu_0 6(2t)^3, \quad -\int\limits_{V(t)} K\, dV \geq K_0 (2t)^3,$$

$$\sum_{k=1}^{3} \int\int\limits_{x_k = \pm t} \frac{\xi_i^2 + \xi_j^2}{2}\, dx_i\, dx_j \leq 12t^2,$$

we obtain

$$3a^2 \geq K_0 \frac{a^4}{8} - \mu_0 a^3.$$

From this the required inequality follows.

We say that *the vector field* $\mathbf{A}$ and the family of orthogonal planes *are algebraic of order m if the components* $\mathbf{A}_i(x_1, x_2, x_3)$ *are the polynomials of order not higher than* m. (Here, the vector $\mathbf{A}$ is not necessarily unit; if we have the algebraic family of surfaces then we get the algebraic family of planes.) It is evident that the algebraic character of a vector field and its order do not depend on the choice of Cartesian coordinates. With the help of formula (1) of Section 1.7 we can prove the following theorem.

Theorem *Let us be given a regular algebraic vector field of order m in a cube V with edge length a. Suppose that the total curvature K satisfies inside V the inequality*

$$|K| \geq K_0 > 0,$$

where K_0 *is constant. Then the edge length is bounded from above, namely*

$$a \leq \frac{Cm}{\sqrt{K_0}}, \quad C = 2\sqrt{3\pi\sqrt{3}}.$$

Here the regularity of the vector field means that $|\mathbf{A}| \neq 0$. We represent the mapping ψ of the boundary ∂V onto a unit sphere by means of the vector field $\mathbf{n}$ as a composition of the mapping φ of ∂V into three-dimensional Euclidean space by means of a vector field $\mathbf{A}$ and the subsequent projection χ onto unit sphere by means of the rays from the origin: $\psi = \chi\varphi$.

Consider the image of the face $x_1 = \text{const}$ under the mapping φ. Denote it by D. A common point of D and the ray from the origin directed as $(\alpha_1, \alpha_2, \alpha_3)$ has coordinates satisfying

$$\frac{A_1}{\alpha_1} = \frac{A_2}{\alpha_2} = \frac{A_3}{\alpha_3}.$$

We may assume that $\alpha_2 \neq 0$, for instance. Set

$$P(x_2, x_3) = \alpha_2 A_1 - \alpha_1 A_2, \quad Q(x_2, x_3) = \alpha_3 A_2 - \alpha_2 A_3.$$

Then at the common point of D and the ray we have $P = Q = 0$ due to the equations above. Since the orders of polynomials P and Q do not exceed m, the number of points in the face $x_1 =$ const such that the Jacobian $\mathbf{I}\left(\frac{P.Q}{x_2.x_3}\right) \neq 0$ and $P = Q = 0$ does not exceeds m^2. Consider the points at which $\mathbf{I}\left(\frac{P.Q}{x_2.x_3}\right) = 0$. At those points we have the following equation:

$$\alpha_2\alpha_1\begin{vmatrix} A_{2x_2} & A_{2x_3} \\ A_{3x_2} & A_{3x_3} \end{vmatrix} + \alpha_2^2\begin{vmatrix} A_{3x_2} & A_{1x_2} \\ A_{3x_3} & A_{1x_3} \end{vmatrix} + \alpha_2\alpha_3\begin{vmatrix} A_{1x_2} & A_{2x_2} \\ A_{1x_3} & A_{2x_3} \end{vmatrix} = 0. \tag{5}$$

Evidently, the normal to D is collinear to the vector

$$\mathbf{I} = \left(\begin{vmatrix} A_{2x_2} & A_{2x_3} \\ A_{3x_2} & A_{3x_3} \end{vmatrix}, \begin{vmatrix} A_{3x_2} & A_{1x_2} \\ A_{3x_3} & A_{1x_3} \end{vmatrix}, \begin{vmatrix} A_{1x_2} & A_{2x_2} \\ A_{1x_3} & A_{2x_3} \end{vmatrix}\right).$$

Therefore, by virtue of (5) either $(\alpha_1, \alpha_2, \alpha_3)$ is tangent to D or $\mathbf{I} = 0$. For the first case the Jacobian of χ is zero, for the second case the Jacobian of φ is zero. By the Sard theorem the set of those points in the image, in our case in the unit sphere, is of zero measure. So, each of the rays from the origin, excluding the set of zero measure, intersect D no more than m^2 times. Now put the origin into the center of the cube V. Then $|(\mathbf{x}, \mathbf{n})| \leq \frac{a\sqrt{3}}{2}$. Hence, using (1) of Section 1.7 , we have

$$K_0 a^3 \leq \left|\int_V K\,dV\right| \leq \left|\int_{\chi\varphi(\partial V)} (\mathbf{x}, \mathbf{n})\,d\sigma\right| \leq 12\sqrt{3}a\pi m^2.$$

The theorem is proved.

The problems considered in this section was brought into being by the ideas of the well-known Efimov theorem on the regular projection of a surface $Z = Z(x, y)$ of negative curvature $K \leq -1$ onto the square in the xy-plane (see [62]). That theorem states that *the edge length a of the square is bounded from above by the independent constant* $a \leq 14$.

That theorem founded Efimov's remarkable investigations of negatively curved surfaces which were awarded Lenin's prize. The central result of those investigations represents the following statement.

Theorem *For any C^2 regular complete surface of negative curvature in Euclidean space E^3 holds* $\inf|K| = 0$ *(see [63]–[65]).*

Thus, the generalization of Hilbert's theorem from the surfaces of constant negative curvature to the surfaces of variable negative curvature was given. That problem interested many famous geometers and was not solved for a long time.

The proof is based on the following lemma, which we cite because of its applications not only to the theory of surfaces but, as Efimov suggested, to the theory of vector fields.

Efimov's Lemma Let us be given a mapping of the whole xy-plane into a pq-plane:

$$p = p(x, y), \quad q = q(x, y),$$

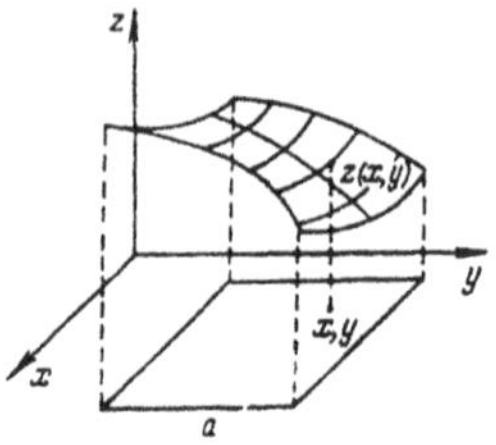

FIGURE 16

where p, q are of class C^1. Suppose that the Jacobian Δ satisfies $\Delta = \begin{vmatrix} p_x & q \\ p_y & q \end{vmatrix}$
and the rotation $I = p_y - q_x$ satisfies the inequality

$$|\Delta| \geq a|I| + a^2$$

where $a > 0$ is some constant. Then the image of the whole xy-plane is eith whole pq-plane or a half-plane or a zone between parallel straight lines; the ma of the xy-plane onto these domains is a homeomorphism.

In the theory of surfaces the lemma is used only when $I = 0$. The value rotation) is analogous to the non-holonomicity value of the vector field. I independent of x_3 and $\xi_3 \neq 0$ then setting $p = \xi_1/\xi_3$, $q = \xi_2/\xi_3$ we obtain th lowing expression for the non-holonomicity value and the curvature K:

$$(\mathbf{n}, \operatorname{curl} \mathbf{n}) = \frac{p_y - q_x}{1 + p^2 + q^2}, \quad K = \frac{\Delta}{(1 + p^2 + q^2)^2}.$$

Therefore, *if the invariants of the field* $\mathbf{n}$ *satisfy the inequality* I $|K| \geq a|(\mathbf{n}, \operatorname{curl} \mathbf{n})| + a^2$ *then the mapping* $p(x, y)$, $q(x, y)$ *satisfies the hypothe Efimov's lemma.*

Examples of mappings satisfying the hypothesis of the lemma and which r plane onto a plane or a half-plane are known. In [92] the mapping onto a half-] $p = \ln(x + \sqrt{x^2 + e^{-2y}}) + y$, $q = \sqrt{x^2 + e^{-2y}}$ has been constructed. *The probl the existence of such mapping onto a zone is still open.*

1.15 The Behavior of Vector Field Streamlines in the Neighborhood of a Close Streamline

Let us be given a unit vector field $\mathbf{n}$ defined in some domain of three-dimens Euclidean space E^3. Let L be a closed streamline of that vector field. Conside following aspects of streamline behavior near L: (1) the rotation of bands form the field $\mathbf{n}$ streamlines with respect to the band of principal normals of L; (2 stability of L as a closed trajectory of the solution, differential equation $d\mathbf{x}/dt =$

where $x \in E^3$. In some cases we can gain information about streamline behavior near L by the geometrical invariants of the field only at L. In considering the problems which are posed one of those invariants arises naturally, namely

$$\Lambda = \sqrt{(\operatorname{div}\mathbf{n})^2 - 4K + (\mathbf{n}, \operatorname{curl}\mathbf{n})^2} \tag{1}$$

It happens that if $\Lambda = 0$ then the streamlines infinitely close to L behave uniformly.

Take a point $P_0 \in L$ and draw a plane F orthogonal to L at P_0. In that plane we take the infinitely small intercept α which passes through P_0 and then draw the field $\mathbf{n}$ streamlines through the points of α (see Fig. 17). The band intersects the normal plane at an arbitrary point P in L by some intercept $\alpha(P)$ which makes some angle φ with the principal normal of L. We suppose that φ is a continuous function of the arc length parameter of L. We also suppose that the field of principal normals of L is continuously and uniquely prolongable. After the first bypass along L the band intersects F at an by infinitely small intercept α_1 which will be twisted in some angle $\Delta\varphi$ with respect to α.

Theorem *Let L be a closed streamline of the field* $\mathbf{n}$. *Suppose that $\Lambda = 0$ at L. Then the twist angle of all bands formed with the streamlines is the same and is equal to*

$$\int_L \left\{\frac{1}{2}(\mathbf{n}, \operatorname{curl}\mathbf{n}) - \kappa\right\} ds, \tag{2}$$

where κ is the torsion of L.

If the integral (2) is a rational number p/q after being divided by 2π then the band will take its starting position after q turns with respect to L and p turns with respect to the band of principal normals; if that number is irrational then the band will wind around L everywhere dense.

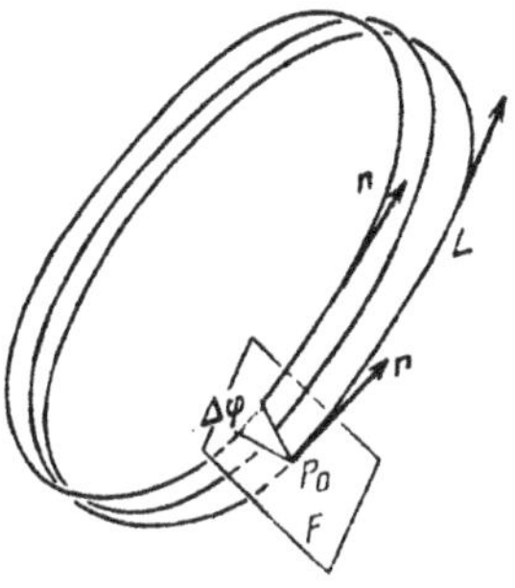

FIGURE 17

Consider the invariant Λ. In the case when $\mathbf{n}$ is holonomic, i.e. it is orthogonal to the family of surfaces with $1/R_i$ as the principal curvature, then

$$\Lambda = \left|\frac{1}{R_1} - \frac{1}{R_2}\right|. \tag{3}$$

In the sequel we shall need another expression for Λ:

$$\Lambda = \sqrt{[(\mathbf{a}, \operatorname{curl}\mathbf{a}) - (\mathbf{b}, \operatorname{curl}\mathbf{b})]^2 + [(\mathbf{a}, \operatorname{curl}\mathbf{b}) + (\mathbf{b}, \operatorname{curl}\mathbf{a})]^2}, \tag{4}$$

where $\mathbf{a}, \mathbf{b}$ are unit mutually orthogonal vectors such that $[\mathbf{a}, \mathbf{b}] = \mathbf{n}$. To prove this, we use the following formula which is valid for any two vector fields $\mathbf{a}$ and $\mathbf{b}$:

$$\operatorname{curl}[\mathbf{a}, \mathbf{b}] = \mathbf{a}\operatorname{div}\mathbf{b} - \mathbf{b}\operatorname{div}\mathbf{a} + \nabla_{\mathbf{b}}\mathbf{a} - \nabla_{\mathbf{a}}\mathbf{b}, \tag{5}$$

where, for instance, $\nabla_{\mathbf{b}}\mathbf{a}$ means the derivative of $\mathbf{a}$ along $\mathbf{b}$. It is possible to write

$$\operatorname{curl}\mathbf{a} = \operatorname{curl}[\mathbf{b}, \mathbf{n}] = \mathbf{b}\operatorname{div}\mathbf{n} - \mathbf{n}\operatorname{div}\mathbf{b} + \nabla_{\mathbf{n}}\mathbf{b} - \nabla_{\mathbf{b}}\mathbf{n},$$

$$\operatorname{curl}\mathbf{b} = \operatorname{curl}[\mathbf{n}, \mathbf{a}] = \mathbf{n}\operatorname{div}\mathbf{a} - \mathbf{a}\operatorname{div}\mathbf{n} + \nabla_{\mathbf{a}}\mathbf{n} - \nabla_{\mathbf{n}}\mathbf{a}.$$

From this we find

$$(\mathbf{a}, \operatorname{curl}\mathbf{a}) = (\mathbf{a}, \nabla_{\mathbf{n}}\mathbf{b} - \nabla_{\mathbf{b}}\mathbf{n}),$$

$$(\mathbf{b}, \operatorname{curl}\mathbf{b}) = (\mathbf{b}, \nabla_{\mathbf{a}}\mathbf{n} - \nabla_{\mathbf{n}}\mathbf{a}).$$

Subtracting the second from the first and taking into account that $(\mathbf{a}, \nabla_{\mathbf{n}}\mathbf{b}) + (\mathbf{b}, \nabla_{\mathbf{n}}\mathbf{a}) = 0$, we obtain

$$(\mathbf{a}, \operatorname{curl}\mathbf{a}) - (\mathbf{b}, \operatorname{curl}\mathbf{b}) = -(\mathbf{a}, \nabla_{\mathbf{b}}\mathbf{n}) - (\mathbf{b}, \nabla_{\mathbf{a}}\mathbf{n}). \tag{6}$$

Multiplying the expression for $\operatorname{curl}\mathbf{a}$ and $\operatorname{curl}\mathbf{b}$ by $\mathbf{b}$ and $\mathbf{a}$ respectively and then adding, we get

$$(\mathbf{a}, \operatorname{curl}\mathbf{b}) + (\mathbf{b}, \operatorname{curl}\mathbf{a}) = (\mathbf{a}, \nabla_{\mathbf{a}}\mathbf{n}) - (\mathbf{b}, \nabla_{\mathbf{b}}\mathbf{n}). \tag{7}$$

From (6) and (7) follows:

$$[(\mathbf{a}, \operatorname{curl}\mathbf{a}) - (\mathbf{b}, \operatorname{curl}\mathbf{b})]^2 + [(\mathbf{a}, \operatorname{curl}\mathbf{b}) + (\mathbf{b}, \operatorname{curl}\mathbf{a})]^2$$
$$= [(\mathbf{a}, \nabla_{\mathbf{b}}\mathbf{n}) + (\mathbf{b}, \nabla_{\mathbf{a}}\mathbf{n})]^2 + [(\mathbf{a}, \nabla_{\mathbf{a}}\mathbf{n}) - (\mathbf{b}, \nabla_{\mathbf{b}}\mathbf{n})]^2.$$

Now choose the Cartesian coordinates x_1, x_2, x_3 in such a way that at some fixed point M the x_3-axis would be directed along $\mathbf{n}(M)$, the x_1-axis along $\mathbf{a}$ and the x_2-axis along $\mathbf{b}$. Suppose that ξ_1, ξ_2, ξ_3 are the components of the field $\mathbf{n}$ with respect to that system of coordinates. Note that at M we have $\xi_1 = \xi_2 = \xi_{3x_i} = 0$, $i = 1, 2, 3$. Then we can write

$$(\mathbf{a}, \nabla_{\mathbf{b}}\mathbf{n}) + (\mathbf{b}, \nabla_{\mathbf{a}}\mathbf{n}) = \xi_{1x_2} + \xi_{2x_1},$$

$$(\mathbf{a}, \nabla_{\mathbf{a}}\mathbf{n}) - (\mathbf{b}, \nabla_{\mathbf{b}}\mathbf{n}) = \xi_{1x_1} - \xi_{2x_2}.$$

From this we find

$$
\begin{aligned}
&[(\mathbf{a}, \nabla_{\mathbf{b}}\mathbf{n}) + (\mathbf{b}, \nabla_{\mathbf{a}}\mathbf{n})]^2 + [(\mathbf{a}, \nabla_{\mathbf{a}}\mathbf{n}) - (\mathbf{b}, \nabla_{\mathbf{b}}\mathbf{n})]^2 \\
&\quad = (\xi_{1x_2} + \xi_{2x_1})^2 + (\xi_{1x_1} - \xi_{2x_2})^2 \\
&\quad = (\xi_{1x_1} + \xi_{2x_2})^2 - 4(\xi_{1x_1}\xi_{2x_2} - \xi_{1x_2}\xi_{2x_1}) + (\xi_{1x_2} - \xi_{2x_1})^2 \\
&\quad = (\operatorname{div}\mathbf{n})^2 - 4K + (\mathbf{n}, \operatorname{curl}\mathbf{n})^2 = \Lambda^2,
\end{aligned}
$$

which proves (4).

Below we shall need the following relation:

$$\operatorname{div}\mathbf{n} = \xi_{1x_1} + \xi_{2x_2} = (\mathbf{a}, \nabla_{\mathbf{a}}\mathbf{n}) + (\mathbf{b}, \nabla_{\mathbf{b}}\mathbf{n}). \tag{8}$$

Lemma 1 Let η, ν be the fields of principal normals and binormals of the field **n** streamlines respectively. Then the torsion κ of the field **n** streamline is

$$\kappa = \frac{1}{2}((\mathbf{n}, \operatorname{curl}\mathbf{n}) - (\eta, \operatorname{curl}\eta) - (\nu, \operatorname{curl}\nu)). \tag{9}$$

Use formula (6). We apply it to the unit vector fields **n**, η and ν. We have

$$\operatorname{curl}\nu = \operatorname{curl}[\mathbf{n}, \eta] = \mathbf{n}\operatorname{div}\eta - \eta\operatorname{div}\mathbf{n} + \nabla_{\eta}\mathbf{n} - \nabla_{\mathbf{n}}\eta. \tag{10}$$

Hence, the non-holonomicity value of the binormal field is

$$(\nu, \operatorname{curl}\nu) = -\kappa + (\nu, \nabla_{\eta}\mathbf{n}). \tag{11}$$

Next, by means of (5), we find

$$\operatorname{curl}\eta = \operatorname{curl}[\nu, \mathbf{n}] = \nu\operatorname{div}\mathbf{n} - \mathbf{n}\operatorname{div}\nu + \nabla_{\mathbf{n}}\nu - \nabla_{\nu}\mathbf{n}. \tag{12}$$

Therefore, the non-holonomicity value of the principal normal field is

$$(\eta, \operatorname{curl}\eta) = -\kappa - (\eta, \nabla_{\nu}\mathbf{n}). \tag{13}$$

Finally, we find the non-holonomicity value of the basic vector field **n**.

$$(\mathbf{n}. \operatorname{curl}\mathbf{n}) = (\mathbf{n}, \nabla_{\nu}\eta) - (\mathbf{n}, \nabla_{\eta}\nu) = (\nu, \nabla_{\eta}\mathbf{n}) - (\eta, \nabla_{\nu}\mathbf{n}). \tag{14}$$

Adding (11) and (13) and then subtracting (14), we obtain the required formula (9).

Lemma 2 Let **a**, **b** be the vector fields such that

$$\mathbf{a} = \cos\varphi\eta + \sin\varphi\nu, \quad \mathbf{b} = -\sin\varphi\eta + \cos\varphi\nu.$$

Then

$$2\kappa = (\mathbf{n}, \operatorname{curl}\mathbf{n}) - (\mathbf{a}, \operatorname{curl}\mathbf{a}) - (\mathbf{b}, \operatorname{curl}\mathbf{b}) - 2\frac{d\varphi}{ds}, \tag{15}$$

where $d\varphi/ds$ is a derivative along the streamline of the field **n**.

By direct calculation we find

$$(\mathbf{a}, \operatorname{curl} \mathbf{a}) = \cos^2\varphi(\eta, \operatorname{curl}\eta) + \sin^2\varphi(\nu, \operatorname{curl}\nu) + \cos\varphi\sin\varphi\{(\nu, \operatorname{curl}\eta) + (\eta, \operatorname{curl}\nu)\} - \frac{d\varphi}{ds}.$$

We optain the corresponding expression for $(\mathbf{b}, \operatorname{curl}\mathbf{b})$ from the previous expression replacing φ with $\varphi + \pi/2$. We have

$$(\mathbf{b}, \operatorname{curl} \mathbf{b}) = \sin^2\varphi(\eta, \operatorname{curl}\eta) + \cos^2\varphi(\nu, \operatorname{curl}\nu) - \cos\varphi\sin\varphi\{(\nu, \operatorname{curl}\eta) + (\eta, \operatorname{curl}\nu)\} - \frac{d\varphi}{ds}.$$

Adding the latter expression we find

$$(\mathbf{a}, \operatorname{curl}\mathbf{a}) + (\mathbf{b}, \operatorname{curl}\mathbf{b}) = (\eta, \operatorname{curl}\eta) + (\nu, \operatorname{curl}\nu) - 2\frac{d\varphi}{ds}.$$

From this and lemma 1, (15) follows.

Now consider a band formed by the streamlines. Evidently, it may be included in a regular family of such bands. Define a vector field $\mathbf{b}$ as a field of normals to those bands. By definition, the field $\mathbf{b}$ is holonomic, i.e. $(\mathbf{b}, \operatorname{curl}\mathbf{b}) = 0$. Take a vector field $\mathbf{a}$ tangent to the band and orthogonal to $\mathbf{n}$. From (15) we find that the turn of such a band with respect to the band of principal normals η is equal to

$$\int_L \left\{\frac{1}{2}(\mathbf{n}, \operatorname{curl}\mathbf{n}) - \kappa - \frac{1}{2}(\mathbf{a}, \operatorname{curl}\mathbf{a})\right\} ds.$$

For an arbitrary field $\mathbf{n}$ the value $(\mathbf{a}, \operatorname{curl}\mathbf{a})$ depends on the band choice. If at L we have $\Lambda = 0$ then by virtue of (4) $(\mathbf{a}, \operatorname{curl}\mathbf{a}) = (\mathbf{b}, \operatorname{curl}\mathbf{b}) = 0$. Therefore, all the bands turn by the same angle. This proves the theorem.

Turn now to the stability problem of the closed streamline. Choose a positive direction in L. We are going to find a condition, in terms of the geometrical invariants along that curve of the field $\mathbf{n}$, under which L will be the orbitally asymptotically stable limit cycle of the differential equation $d\mathbf{x}/dt = \mathbf{n}(x)$.

The stability condition below is a generalization of the Poincaré condition for the field in a plane and expresses the Vagewski stability criterion in geometrical terms.

Theorem *Let L be a closed trajectory of the differential equation $d\mathbf{x}/dt = \mathbf{n}(\mathbf{x})$. If*

$$\int_L \operatorname{div}\mathbf{n}\, ds < -\int_L \Lambda\, ds,$$

then L is an orbitally stable limit cycle.

The integral is assumed in the positive direction of L which corresponds to the positive direction of t. Observe that if $\Lambda = 0$ at L then all the nearby trajectories behave uniformly with respect to the stability property.

Let $\mathbf{r}^0(s)$ be a position vector of L, s its arc length parameter, l the length of the curve. Suppose that $\mathbf{a}$ and $\mathbf{b}$ are the two mutually orthogonal unit vector fields which are orthogonal to **n** and periodic at L. Let $P(s)$ be a point in L and $F(s)$ be a normal to the L plane at this point. In the plane $F(0)$ we take the intercept α passing through $P_0 = P(0)$ and emit the field **n** streamlines from the points of α. Let P_0' be a point in α and $\varepsilon_0 = |P_0P_0'|$. Denote by L' a streamline emitted from P_0' and infinitely close to L. Let $P'(s)$ be a common point of L' and $F(s)$. Then we set $\varepsilon(s) = PP'$. We represent a position vector infinitely close to the L trajectory, namely L', as

$$\mathbf{r}(s) = \mathbf{r}^0(s) + \varepsilon(s, \varepsilon_0)\{\cos\varphi(s, \varepsilon_0)\mathbf{a}(s) + \sin\varphi(s, \varepsilon_0)\mathbf{b}(s)\}.$$

Denote the arc length parameter of L' by u. We have

$$\mathbf{r}_u \frac{du}{ds} = \mathbf{r}_s^0 + \varepsilon_s(\cos\varphi\,\mathbf{a} + \sin\varphi\,\mathbf{b}) + \varepsilon(\cos\varphi\,\mathbf{a} + \sin\varphi\,\mathbf{b})_s.$$

The vector $\mathbf{r}_u$ coincides with $\mathbf{n}(P')$. Multiply it by $(\cos\varphi\,\mathbf{a} + \sin\varphi\,\mathbf{b})$. Then

$$\frac{d\varepsilon(u, \varepsilon_0)}{du} = (\mathbf{n}(P'), \cos\varphi\,\mathbf{a} + \sin\varphi\,\mathbf{b}). \tag{17}$$

Consider the expansion

$$\mathbf{n}(P') = \mathbf{n}(P) + \varepsilon(u, \varepsilon_0)(\cos\varphi\,\nabla_{\mathbf{a}}\mathbf{n} + \sin\varphi\,\nabla_{\mathbf{b}}\mathbf{n}) + o(\varepsilon).$$

Set $\frac{d\varepsilon(s,0)}{d\varepsilon_0} = f$. Substitute $\mathbf{n}(P')$ into (17), divide both sides by ε_0 and then pass to the limit with respect to ε_0. We consider only one loop of L'. It is possible to show that there exists $\lim\varphi(s, \varepsilon_0)$ when $\varepsilon_0 \to 0$. Denote this limit by $\varphi(s)$. Then at the points of L we obtain

$$\begin{aligned}\frac{d\ln|f(s)|}{ds} &= (\nabla_{\mathbf{a}}\mathbf{n}, \mathbf{a})\cos^2\varphi + (\nabla_{\mathbf{b}}\mathbf{n}, \mathbf{b})\sin^2\varphi + \cos\varphi\,\sin\varphi\,[(\nabla_{\mathbf{a}}\mathbf{n}, \mathbf{b}) + (\nabla_{\mathbf{b}}\mathbf{n}, \mathbf{a})]\\ &= \frac{1}{2}[(\nabla_{\mathbf{a}}\mathbf{n}, \mathbf{a}) + (\nabla_{\mathbf{b}}\mathbf{n}, \mathbf{b})] + \frac{1}{2}\cos 2\varphi\,[(\nabla_{\mathbf{a}}\mathbf{n}, \mathbf{a}) - (\nabla_{\mathbf{b}}\mathbf{n}, \mathbf{b})]\\ &\quad + \frac{1}{2}\sin 2\varphi[(\nabla_{\mathbf{a}}\mathbf{n}, \mathbf{b}) + (\nabla_{\mathbf{b}}\mathbf{n}, \mathbf{a})].\end{aligned}$$

Use formulas (8), (7) and (6):

$$\begin{aligned}\frac{d\ln|f(s)|}{ds} &= \frac{1}{2}\operatorname{div}\mathbf{n} - \frac{1}{2}\sin 2\varphi[(\mathbf{a}, \operatorname{curl}\mathbf{a}) - (\mathbf{b}, \operatorname{curl}\mathbf{b})]\\ &\quad + \frac{1}{2}\cos 2\varphi[(\mathbf{a}, \operatorname{curl}\mathbf{b}) + (\mathbf{b}, \operatorname{curl}\mathbf{a})].\end{aligned} \tag{18}$$

Note that by virtue of (4) the sum of the two last terms is less than $\Lambda(s)$. Integrating (18) along L in the positive direction and using condition (16), we see that $\left|\frac{f(l)}{f(0)}\right| < 1$ uniformly for any starting intercept α in the plane $F(0)$. This inequality is sufficient for stability.

It is easy to produce an example of a field with a closed stable trajectory L satisfying the hypothesis of the theorem. Take any closed spatial curve L and draw through any point $P \in L$ a piece of surface of strictly positive curvature turned with its concavity to the positive direction of L and perpendicular to L. Let R_1 and R_2 be the principal curvatures of that surface. In moving P along L we obtain the regular family of surfaces in some neighborhood of L. Define **n** as a field of normals of that family. In this case $\Lambda = \left|\frac{1}{R_1} - \frac{1}{R_2}\right|$, $\operatorname{div}\mathbf{n} = -\left(\frac{1}{R_1} + \frac{1}{R_2}\right) < -\Lambda$. Hence, L is a stable limit cycle. If at least one of the principal curvatures is zero then the closed trajectories occur, maybe, near L, i.e. L cannot be the limit cycle.

1.16 The Complex Non-Holonomicity

We shall introduce a new geometrical object connected with the behavior of the streamlines for the given field **n** in a domain $G \subset E^3$. Let **a**, **b** be the unit mutually orthogonal vectors such that $[\mathbf{a}, \mathbf{b}] = \mathbf{n}$. Consider the field of complex vectors in G:

$$\beta = \mathbf{a} + i\,\mathbf{b}.$$

It is natural to define $\operatorname{curl}\beta$ as follows:

$$\operatorname{curl}\beta = \operatorname{curl}\mathbf{a} + i\operatorname{curl}\mathbf{b}.$$

The complex non-holonomicity value is a complex number:

$$(\beta, \operatorname{curl}\beta) = (\mathbf{a}, \operatorname{curl}\mathbf{a}) - (\mathbf{b}, \operatorname{curl}\mathbf{b}) + i\{(\mathbf{b}, \operatorname{curl}\mathbf{a}) + (\mathbf{a}, \operatorname{curl}\mathbf{b})\}.$$

The modulus of this number does not depend on the choice of **a** and **b** (see Section 1.14), i.e. it *is the invariant of the field* **n** *and is equal to* Λ:

$$|(\beta, \operatorname{curl}\beta)|^2 = (\operatorname{div}\mathbf{n})^2 - 4K + (\mathbf{n}, \operatorname{curl}\mathbf{n})^2 = \Lambda^2.$$

In rotating **a** and **b** in the plane orthogonal to **n** the value $(\beta, \operatorname{curl}\beta)$ changes in a special manner. Let **a**′, **b**′ be the vector fields obtained from **a** and **b** as a result of rotation by the angle $\varphi = \varphi(x_1, x_2, x_3)$:

$$\mathbf{a}' = \cos\varphi\,\mathbf{a} + \sin\varphi\,\mathbf{b},$$
$$\mathbf{b}' = -\sin\varphi\,\mathbf{a} + \cos\varphi\,\mathbf{b},$$

i.e. $\beta' = e^{-i\varphi}\beta$. We have

$$\operatorname{curl}\beta' = e^{-i\varphi}(\operatorname{curl}\beta - i[\operatorname{grad}\varphi, \beta]).$$

Hence, the complex non-holonomicity of the field β' has the form

$$(\beta', \operatorname{curl}\beta') = e^{-2i\varphi}\{(\beta, \operatorname{curl}\beta) - i(\beta\operatorname{grad}\varphi, \beta)\}$$
$$= e^{-2i\varphi}(\beta, \operatorname{curl}\beta),$$

because of $(\beta, \operatorname{grad}\varphi, \beta) \equiv 0$.

We shall say that a complex vector field β is *holonomic* if there are complex functions $\lambda(x_1, x_2, x_3)$ and $\psi(x_1, x_2, x_3)$ of real variables x_1, x_2, x_3 in a domain $G \subset E^3$ such that

$$\beta = \lambda\operatorname{grad}\psi. \tag{1}$$

Theorem *A complex vector field β is holonomic if and only if the complex non-holonomicity value is zero:*

$$(\beta, \operatorname{curl}\beta) = 0. \tag{2}$$

If (1) is satisfied then

$$\begin{aligned}(\beta, \operatorname{curl}\beta) &= \lambda(\operatorname{grad}\psi, \lambda \operatorname{curl}\operatorname{grad}\psi + [\operatorname{grad}\lambda, \operatorname{grad}\psi]) \\ &= \lambda(\operatorname{grad}\psi, \operatorname{grad}\lambda, \operatorname{grad}\psi) = 0,\end{aligned}$$

i.e. (2) is satisfied, too. Conversely, suppose that $(\beta, \operatorname{curl}\beta) = 0$. Show that every vector field $\beta = \mathbf{a} + i\mathbf{b}$ generated by the vector fields $\mathbf{a}$ and $\mathbf{b}$ and orthogonal to $\mathbf{n}$ is holonomic.

Find, first, a single holonomic vector field β. To do this we rewrite the holonomicity condition in terms of real functions. Let $\lambda = \lambda_1 + i\lambda_2$, $\psi = \psi_1 + i\psi_2$. Then (1) can be expanded to the system of two differential equations:

$$\begin{aligned}\mathbf{a} &= \lambda_1 \operatorname{grad}\psi_1 - \lambda_2 \operatorname{grad}\psi_2, \\ \mathbf{b} &= \lambda_2 \operatorname{grad}\psi_1 + \lambda_1 \operatorname{grad}\psi_2.\end{aligned} \tag{3}$$

We draw a surface F through the point $P_0 \in G$ such that in a sufficiently small neighborhood of P_0 the field $\mathbf{n}$ restricted on F is not tangent to F. We define a regular family of curves in that neighborhood restricted on F. Denote it by γ_1. Then we draw a streamline of the field $\mathbf{n}$ through the points of each curve. We obtain a family of surfaces $\varphi_1(x_1, x_2, x_3) = \text{const}$ made up from the field $\mathbf{n}$ streamlines. Let $\mathbf{a}$ be a unit normal field of that family of surfaces, i.e. $\mathbf{a} = \lambda_1 \operatorname{grad}\varphi_1$. Since this field is holonomic by construction, $(\mathbf{a}, \operatorname{curl}\mathbf{a}) = 0$. Set $\mathbf{b} = [\mathbf{n}, \mathbf{a}]$. By the hypothesis $(\mathbf{b}, \operatorname{curl}\mathbf{b}) = (\mathbf{a}, \operatorname{curl}\mathbf{a}) = 0$. Hence, the field $\mathbf{b}$ is also holonomic, i.e. there exists a family of surfaces $\varphi_2(x_1, x_2, x_3) = \text{const}$ such that

$$\begin{aligned}\mathbf{a} &= \lambda_1 \operatorname{grad}\varphi_1, \\ \mathbf{b} &= \lambda_2 \operatorname{grad}\varphi_2.\end{aligned}$$

Since $\mathbf{n}$ is orthogonal to $\mathbf{b}$, $\mathbf{n}$ is tangent to the surfaces $\varphi_2 = \text{const}$. Hence, those surfaces are formed with the field $\mathbf{n}$ streamlines and the surfaces from different families, namely $\varphi_1 = \text{const}$ and $\varphi_2 = \text{const}$, intersect each other by a streamline. From the condition $(\mathbf{a}, \operatorname{curl}\mathbf{b}) + (\mathbf{b}, \operatorname{curl}\mathbf{a}) = 0$ the relation for λ_1 and λ_2 follows. Since

$$\begin{aligned}(\mathbf{a}, \operatorname{curl}\mathbf{b}) &= \lambda_1(\operatorname{grad}\varphi_1, \operatorname{grad}\lambda_2, \operatorname{grad}\varphi_2), \\ (\mathbf{b}, \operatorname{curl}\mathbf{a}) &= \lambda_2(\operatorname{grad}\varphi_2, \operatorname{grad}\lambda_1, \operatorname{grad}\varphi_1),\end{aligned}$$

it follows that

$$(\mathbf{a}, \operatorname{curl}\mathbf{b}) + (\mathbf{b}, \operatorname{curl}\mathbf{a}) = (\lambda_1^2 + \lambda_2^2)\left(\operatorname{grad}\varphi_1, \operatorname{grad}\varphi_2, \operatorname{grad}\arctan\frac{\lambda_1}{\lambda_2}\right) = 0.$$

The gradient of λ_1/λ_2 is a linear combination of $\operatorname{grad}\varphi_1$ and $\operatorname{grad}\varphi_2$ and hence, is orthogonal to the streamlines of the field $\mathbf{n}$. Therefore, the surfaces $\lambda_1/\lambda_2 = \text{const}$ are formed by the streamlines of the field **n**, i.e. those surfaces have the same structure as $\varphi_1 = \text{const}$ and $\varphi_2 = \text{const}$. Namely, those surfaces intersect the surface F by some families of curves γ_1 and γ_2 and the corresponding field streamlines are drawn through the points of the curves of the families. By means of the families γ_1 and γ_2, we introduce curvilinear coordinates φ_1 and φ_2 such that the curves $\varphi_i = \text{const}$ coincide with the curves γ_i. Assume that the surfaces $\lambda_1/\lambda_2 = \text{const}$ intersect F by the family of curves $\Phi(\varphi_1, \varphi_2) = \text{const}$. Then λ_1/λ_2 is a function of φ_1, φ_2:

$$\frac{\lambda_1}{\lambda_2} = \Phi(\varphi_1, \varphi_2).$$

Therefore, for the vector fields **a**, **b** we have

$$\mathbf{a} = \lambda_2 \Phi(\varphi_1, \varphi_2) \operatorname{grad} \varphi_1,$$
$$\mathbf{b} = \lambda_2 \operatorname{grad} \varphi_2.$$

Choose the functions $\psi_1(\varphi_1, \varphi_2)$ and $\psi_2(\varphi_1, \varphi_2)$ such that **a** and **b** will have the form (3):

$$\mathbf{a} = \lambda_2 \Phi(\varphi_{1_{\psi_1}} \operatorname{grad} \psi_1 + \varphi_{1_{\psi_2}} \operatorname{grad} \psi_2),$$
$$\mathbf{b} = \lambda_2(\varphi_{2_{\psi_1}} \operatorname{grad} \psi_1 + \varphi_{2_{\psi_2}} \operatorname{grad} \psi_2).$$

The following system must be satisfied:

$$\Phi \varphi_{1_{\psi_1}} = \varphi_{2_{\psi_2}},$$
$$\varphi_{2_{\psi_1}} = -\varphi_{1_{\psi_2}} \Phi.$$

Supposing that the Jacobian of change φ_i by ψ_i is non-zero, we can represent the latter system as a system with respect to $\psi_i(\varphi_1, \varphi_2)$:

$$\begin{aligned} \Phi(\varphi_1, \varphi_2) \frac{\partial \psi_2}{\partial \varphi_2} &= \frac{\partial \psi_1}{\partial \varphi_1}, \\ -\Phi(\varphi_1, \varphi_2) \frac{\partial \psi_1}{\partial \varphi_2} &= \frac{\partial \psi_2}{\partial \varphi_1}, \end{aligned} \tag{4}$$

Excluding one of the functions, say φ_2, we come to a single equality of elliptic type:

$$\frac{\partial}{\partial \varphi_1}\left(\frac{\partial \psi_1 / \partial \varphi_1}{\Phi}\right) + \frac{\partial}{\partial \varphi_2}\left(\Phi \frac{\partial \psi_1}{\partial \varphi_2}\right) = 0.$$

Find a local solution of this equation such that $\left(\frac{\partial \psi_1}{\partial \varphi_1}\right)^2 + \left(\frac{\partial \psi_1}{\partial \varphi_2}\right)^2 \neq 0$ in the neighborhood of a given point. Then the Jacobian of change φ_i by ψ_i will be non-zero. So, the fields **a**, **b** just constructed, being orthogonal to **n**, are represented in the form (3). The field $\beta = \mathbf{a} + i\mathbf{b}$ is holonomic: $\beta = \lambda \operatorname{grad} \psi$. Any other field $\beta' = e^{-i\varphi}\beta = e^{-i\varphi}\lambda \operatorname{grad} \psi$ is also holonomic. The theorem is proved.

If the field **n** *is holonomic then the complex value of the non-holonomicity value* $(\beta, \operatorname{curl} \beta)$ *is zero if and only if the field* **n** *is a field of normals to the family of spheres or planes.* Indeed, from the expression for Λ it follows that $H^2 = 4K$, i.e. the surfaces orthogonal to **n** are totally umbilical, i.e. they are spheres or planes.

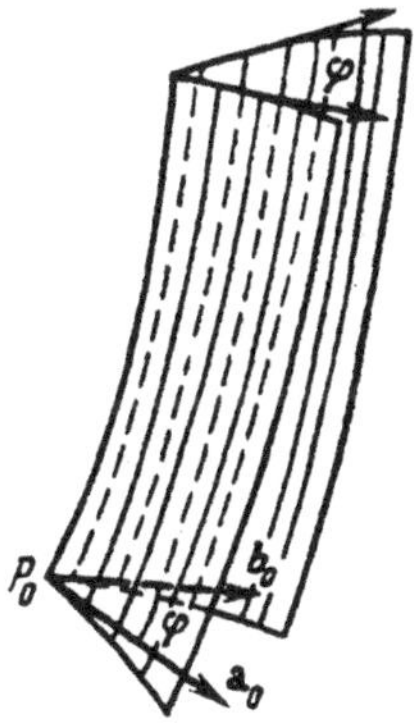

FIGURE 18

The geometrical property of streamlines of the field **n** with the complex value of the non-holonomicity $(\beta, \operatorname{curl} \beta) = 0$ is as follows. Consider a point P_0 and draw a plane having $\mathbf{n}(P_0)$ as the normal. Take the intercepts $\mathbf{a}_0$ and $\mathbf{b}_0$ in that plane having P_0 as their common end-point. Draw the streamlines of the field **n** through each point of the intercepts. So, we obtain two bands of streamlines. The angle between those bands stays constant along their common streamline (see Fig. 18). Indeed, as we have stated in Section 1.15, if $\Lambda = 0$ then all the streamline bands turn in the same angle with respect to the band of principal normals. Hence, the angle between them stays constant.

1.17 The Analogues of Gauss–Weingarten Decompositions and the Bonnet Theorem Analogue

Suppose that besides **n**, the unit mutually orthogonal and orthogonal to **n** vector fields $\mathbf{a}_1, \mathbf{a}_2$ in a domain G are given. Then $\mathbf{a}_1, \mathbf{a}_2$ and **n** form the moving basis in space. We can describe the behavior of the vector field **n** in terms of coefficients in derivation formulas which are analogous to the Gauss–Weingarten decompositions of the surface. Set the indices i, j to be distinct. Consider the decompositions of the field $\mathbf{a}_i$ and **n** derivatives in Cartesian coordinates x_k with respect to the basis $\mathbf{a}_1, \mathbf{a}_2, \mathbf{n}$:

$$\frac{\partial \mathbf{a}_i}{\partial x_k} = \Gamma_{ik} \mathbf{a}_j + N_{ik} \mathbf{n},$$

$$\frac{\partial \mathbf{n}}{\partial x_k} = M_{1k} \mathbf{a}_1 + M_{2k} \mathbf{a}_2.$$

From the conditions $(\mathbf{a}_1, \mathbf{a}_2) = 0$, $(\mathbf{a}_i, \mathbf{n}) = 0$ it follows that $\Gamma_{1k} = -\Gamma_{2k}$, $M_{ik} = -N_{ik}$. Therefore, the system of derivative formulas have the form

$$\frac{\partial \mathbf{a}_i}{\partial x_k} = \Gamma_{ik}\mathbf{a}_j + N_{ik}\mathbf{n}, \tag{1}$$

$$\frac{\partial \mathbf{n}}{\partial x_k} = -N_{1k}\mathbf{a}_1 - N_{2k}\mathbf{a}_2. \tag{2}$$

We call Γ_{ik} *the connection coefficients* and N_{ik} *the second fundamental form coefficients*.

At each point of G three Γ_{ik} and six N_{ik} are defined. Let us find the differential equations that these values satisfy. Differentiating (1) in x_l and using (1), (2), we obtain

$$\frac{\partial^2 \mathbf{a}_i}{\partial x_k \partial x_l} = \left(\frac{\partial \Gamma_{ik}}{\partial x_l} - N_{ik}N_{jl}\right)\mathbf{a}_i + (\Gamma_{ik}\Gamma_{jl} - N_{ik}N_{il})\mathbf{a}_i + \left(\frac{\partial N_{ik}}{\partial x_l} + \Gamma_{ik}N_{jl}\right)\mathbf{n}.$$

Interchanging the roles of k and l, we obtain the expression for $\frac{\partial^2 \mathbf{a}_i}{\partial x_l \partial x_k}$. As x_k mean the Cartesian coordinates in E^3, the second derivatives of $\mathbf{a}_i$ do not depend on the order of differentiation. Equating the expressions for corresponding second derivatives, we obtain the equalities on coefficients of $\mathbf{a}_j$ and $\mathbf{n}$:

$$\frac{\partial \Gamma_{ik}}{\partial x_l} - \frac{\partial \Gamma_{il}}{\partial x_k} + N_{il}N_{jk} - N_{ik}N_{jl} = 0, \tag{3}$$

$$\frac{\partial N_{ik}}{\partial x_l} - \frac{\partial N_{il}}{\partial x_k} + \Gamma_{ik}N_{jl} - \Gamma_{il}N_{jk} = 0. \tag{4}$$

By virtue of $\Gamma_{1k} = -\Gamma_{2k}$ the equality for the coefficients of $\mathbf{a}_i$ holds automatically. If we write the equality condition on mixed second derivatives of $\mathbf{n}$ then we arrive at equation (4). The values Γ_{2k} can be completely excluded from those equations. Therefore, the system (3), (4) consists of three differential equations with derivatives of Γ_{ik} and of six differential equations with derivatives N_{ik}.

The following is the analogue to the Bonnet theorem.

Let us be given for $G \subset E^3$ the functions A_{ik} and B_{ik} of Cartesian coordinates, where $A_{1k} = -A_{2k}$. Suppose that the functions satisfy the system (3), (4) when we replace A_{ik} instead of Γ_{ik} and B_{ik} instead of N_{ik}. Then there are the unit vector fields $\mathbf{a}_1, \mathbf{a}_2, \mathbf{n}$ in G with A_{ik} as the connection coefficients and B_{ik} as the second fundamental form coefficients. The vector fields $\mathbf{a}_1, \mathbf{a}_2, \mathbf{n}$ are defined uniquely up to their choice at one point.

Indeed, the system (3), (4) is a compatibility condition for the system (1), (2) having the fields $\mathbf{a}_1, \mathbf{a}_2, \mathbf{n}$ as a solution. If (3), (4) are satisfied then the solution of (1), (2) exists. Moreover, if at the initial point the triple $\mathbf{a}_1, \mathbf{a}_2, \mathbf{n}$ is orthonormal then it is easy to verify that the solution is orthonormal in G because of the nature of the system.

The set values Γ_{1k} and N_{ik} characterize not only the field $\mathbf{n}$ but the fields $\mathbf{a}_1, \mathbf{a}_2$ also. We would like to obtain some invariants of the field $\mathbf{n}$ itself. Consider the vector fields $\mathbf{b}_1, \mathbf{b}_2$ obtained from $\mathbf{a}_1, \mathbf{a}_2$ by rotation by the angle φ as follows:

$$\mathbf{b}_1 = \cos\varphi\, \mathbf{a}_1 + \sin\varphi\, \mathbf{a}_2,$$
$$\mathbf{b}_2 = -\sin\varphi\, \mathbf{a}_1 + \cos\varphi\, \mathbf{a}_2.$$

Let us have

$$\mathbf{b}_{ix_k} = \bar{\Gamma}_{ik}\mathbf{b}_j + \bar{N}_{ik}\mathbf{n},$$
$$\mathbf{n}_{x_k} = -\bar{N}_{1k}\mathbf{b}_1 - \bar{N}_{2k}\mathbf{b}_2.$$

The new coefficients of that decomposition can be expressed in terms of the old ones:

$$\bar{\Gamma}_{1k} = \Gamma_{1k} + \frac{\partial\varphi}{\partial x_k}, \tag{5}$$

$$\begin{aligned} \bar{N}_{1k} &= N_{1k}\cos\varphi + N_{2k}\sin\varphi, \\ \bar{N}_{2k} &= -N_{1k}\sin\varphi + N_{2k}\cos\varphi. \end{aligned} \tag{6}$$

Hence, the values Γ_{1k} are defined by the field $\mathbf{n}$ up to the derivatives of some function φ in a fixed Cartesian coordinate system. We can regard the set of Γ_{1k} as some vector field $\mathbf{\Gamma}$ in G. From (5) it follows that

$$\operatorname{curl}\bar{\mathbf{\Gamma}} = \operatorname{curl}\mathbf{\Gamma}, \tag{7}$$

Conversely, if (7) holds then $\bar{\mathbf{\Gamma}}$ and $\mathbf{\Gamma}$ differ by the gradient of some function.

Thus, the vector field $\mathbf{\Gamma}$ is the invariant of the field $\mathbf{n}$. Equations (3) produce the expressions of $\operatorname{curl}\mathbf{\Gamma}$ components in terms of N_{ik}. If we introduce two more vector fields $\mathbf{N}_1$ and $\mathbf{N}_2$ of components $\{N_{ik}\}$, $i = 1, 2$ then (3), (4) may be rewritten as

$$\operatorname{curl}\mathbf{\Gamma} = -[\mathbf{N}_1, \mathbf{N}_2],$$
$$\operatorname{curl}\mathbf{N}_1 = -[\mathbf{N}_2, \mathbf{\Gamma}],$$
$$\operatorname{curl}\mathbf{N}_2 = -[\mathbf{\Gamma}, \mathbf{N}_1].$$

Denote the complex vector $\mathbf{N}_1 + i\mathbf{N}_2$ by $\mathbf{N}$. From (6) it follows that the modulus of the field $\mathbf{N}$, i.e. $\sqrt{\mathbf{N}_1^2 + \mathbf{N}_2^2}$, is the invariant of the field $\mathbf{n}$.

1.18 The Triorthogonal Family of Surfaces

Suppose that three surfaces each of which belongs to one of the three families of surfaces defined as $u_1 = \text{const}$, $u_2 = \text{const}$, $u_3 = \text{const}$, where $u = u_i(x_1, x_2, x_3)$ pass through any point of a domain $G \subset E^3$. If they meet each other at right angles then we say that a *triorthogonal family of surfaces* is defined in G (see Fig. 19). Various triorthogonal families are used in the construction of coordinate systems in E^3. We are going to state the famous Dupin theorem on those families.

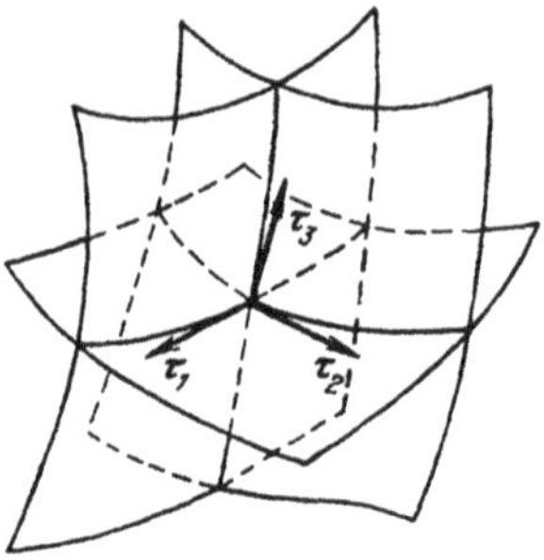

FIGURE 19

Theorem *The surfaces of a triorthogonal family intersect each other by the l*
curvature.

Let τ_i, $i = 1, 2, 3$ be the unit fields tangent to the lines of intersections. Th
have

$$(\tau_1, \tau_2) = 0, \quad (\tau_2, \tau_3) = 0, \quad (\tau_3, \tau_1) = 0.$$

Differentiate the first, the second and the third equation in the directions of
and τ_2 respectively. Then we get

$$(\nabla_{\tau_3}\tau_1, \tau_2) + (\tau_1, \nabla_{\tau_3}\tau_2) = 0,$$
$$(\tau_3, \nabla_{\tau_1}\tau_2) + (\tau_2, \nabla_{\tau_1}\tau_3) = 0,$$
$$(\nabla_{\tau_2}\tau_3, \tau_1) + (\tau_3, \nabla_{\tau_2}\tau_1) = 0.$$

From the first equation of (1) we subtract the second and group together the

$$(\nabla_{\tau_3}\tau_1 - \nabla_{\tau_1}\tau_3, \tau_2) + (\tau_1, \nabla_{\tau_3}\tau_2) - (\tau_3, \nabla_{\tau_1}\tau_2) = 0.$$

Since τ_2 is holonomic and τ_1, τ_3 are orthogonal to τ_2, $\nabla_{\tau_3}\tau_1 - \nabla_{\tau_1}\tau_3$ is orthogoı
τ_2. Hence, (2) implies

$$(\tau_1, \nabla_{\tau_3}\tau_2) = (\tau_3, \nabla_{\tau_1}\tau_2).$$

In an analogous way we obtain two more equations

$$(\tau_2, \nabla_{\tau_3}\tau_1) = (\tau_3, \nabla_{\tau_2}\tau_1),$$
$$(\tau_1, \nabla_{\tau_2}\tau_3) = (\tau_2, \nabla_{\tau_1}\tau_3).$$

The left-hand sides of (3)–(5) are equal to each other. Indeed, consider the
holonomicity condition

$$\nabla_{\tau_3}\tau_2 = \nabla_{\tau_2}\tau_3 + \alpha\tau_2 + \beta\tau_3.$$

Hence,

$$(\tau_1, \nabla_{\tau_1}\tau_2) = (\tau_1, \nabla_{\tau_2}\tau_3). \tag{6}$$

From (5) and (6) we see that the left-hand sides of (3) and (5) are equal to each other. Both sides of (3)–(5) are the scalar products of some τ_k and the derivative of another field τ_l with respect to the third field τ_m, where $k \neq l \neq m \neq k$. So, such scalar products are equal to each other. From (1) we see that all of them are equal to zero. We have, for instance,

$$(\tau_2, \nabla_{\tau_1}\tau_3) = 0. \tag{7}$$

Since τ_3 is a unit field,

$$\nabla_{\tau_1}\tau_3 = \lambda\tau_1,$$

i.e. τ_1 is principal on the surface which is orthogonal to τ_3. In the same way we can prove that τ_1 is principal on the surface which is orthogonal to τ_2. Thus, the line of intersection of those surfaces is the line of curvature with respect to both of them. The theorem is proved.

If we have two families which meet each other at right angles at the lines of curvature then we can complete the triorthogonal family.

Let $u_1 = \text{const}$ and $u_2 = \text{const}$ be given surface families; τ_1, τ_2 be their normal vector fields respectively; τ_3 be the field of tangents to their lines of intersection. We assert that τ_3 is holonomic.

Since τ_3 is a principal direction on the surface $u_2 = \text{const}$, then τ_1 being orthogonal to τ_3 but tangent to the surface is principal, too. Hence,

$$\nabla_{\tau_1}\tau_2 = \mu_1\tau_1.$$

At the same time, τ_2 is principal on the surface $u_1 = \text{const}$ and hence,

$$\nabla_{\tau_2}\tau_1 = \lambda_2\tau_2.$$

So, we have

$$\nabla_{\tau_1}\tau_2 - \nabla_{\tau_2}\tau_1 = \mu_1\tau_1 - \lambda_2\tau_2.$$

Therefore, τ_3 is holonomic: there is a family of orthogonal τ_3 surfaces with $u_3 = \text{const}$.

The direction τ_2 is tangent to the surface $u_1 = \text{const}$ and to the surface $u_3 = \text{const}$ also, because $(\tau_2, \tau_3) = 0$. Therefore, τ_2 is tangent to the line of intersection of $u_1 = \text{const}$ and $u_3 = \text{const}$. Analogously, τ_1 is tangent to the line of intersection of $u_2 = \text{const}$ and $u_3 = \text{const}$. Hence, the new family completes the given two to the triorthogonal family.

If we have only one family given, say $u(x_1, x_2, x_3) = 0$, then it is natural to ask the question: whether or not to complete this family to the triorthogonal one? If this is possible then both of the fields of principal directions are necessarily holonomic, which does not hold for an arbitrary family of surfaces. *In order that the family* $u(x_1, x_2, x_3) = \text{const}$ *be complemented to the triorthogonal family it is necessary and sufficient that the function u satisfies some differential equation. Such a family is called the Lamé family.*

Let $\tau_1 = \{\xi_i\}$, $\tau_3 = \{\zeta_i\}$. Then from (7)

$$0 = (\tau_3 \tau_1 \tau_{3.i})\xi_i = \begin{vmatrix} \zeta_1 & \xi_1 & \zeta_{1.i}\xi_i \\ \zeta_2 & \xi_2 & \zeta_{2.i}\xi_i \\ \zeta_3 & \xi_3 & \zeta_{3.i}\xi_i \end{vmatrix}$$

$$= \varepsilon^{klm} \zeta_k \, \xi_l \, \zeta_{m,i} \, \xi_i = \sum_{l,i} B_{li} \, \xi_l \, \xi_i,$$

where $B_{li} = \varepsilon^{klm} \zeta_k \zeta_{m,i}$. Consider the system of homogeneous equations with respect to ξ_i:

$$\sum_{l,i} B_{li} \, \xi_l \, \xi_i = 0, \tag{8}$$

$$\sum_{i} \zeta_i \, \xi_i = 0. \tag{9}$$

From this system we find two solutions for τ_1:

$$\xi_i = \lambda F_i,$$

where the functions $F_i = F_i(\zeta_k, \zeta_{l,j})$ depend on τ_3, λ is some multiplier. Since τ_1 is holonomic, F_i satisfy the equation

$$F_1 \left(\frac{\partial F_2}{\partial x_3} - \frac{\partial F_3}{\partial x_2} \right) + F_2 \left(\frac{\partial F_3}{\partial x_1} - \frac{\partial F_1}{\partial x_3} \right) + F_3 \left(\frac{\partial F_1}{\partial x_2} - \frac{\partial F_2}{\partial x_1} \right) = 0. \tag{10}$$

Substituting here the expression for F_i in terms of ζ_j and $\zeta_{l,j}$, we obtain the equation which must be satisfied for components of the normal field of a Lamé family.

Consider the system (8)–(10). We can represent equation (8) as

$$\xi_1 \begin{vmatrix} \zeta_2 & \zeta_{2.i} \\ \zeta_3 & \zeta_{3.i} \end{vmatrix} \xi_i - \xi_2 \begin{vmatrix} \zeta_1 & \zeta_{1.i} \\ \zeta_3 & \zeta_{3.i} \end{vmatrix} \xi_i + \xi_3 \begin{vmatrix} \zeta_1 & \zeta_{1.i} \\ \zeta_2 & \zeta_{2.i} \end{vmatrix} \xi_i = 0. \tag{11}$$

Rotate the coordinate system axes in such a way that at a fixed point P_0 the following would be satisfied: $\zeta_1 = \zeta_2 = 0$, $\zeta_3 = 1$. We can regard equation (11) as the equation of some surface in the space of coordinates (ξ_1, ζ_2, ξ_3). At P_0 the equation (11) has the form

$$-\xi_1^2 \zeta_{2.1} + \xi_1 \xi_2 (\zeta_{1.1} - \zeta_{2.2}) - \xi_1 \xi_3 \zeta_{2.3} + \xi_2^2 \zeta_{1.2} + \xi_2 \xi_3 \zeta_{1.3} = 0. \tag{12}$$

Form the matrix B with coefficients of the homogeneous polynomial in the left-hand side of (12):

$$B = \begin{Vmatrix} -\zeta_{2.1} & \frac{\zeta_{1.1} - \zeta_{2.2}}{2} & \frac{-\zeta_{2.3}}{2} \\ \frac{\zeta_{1.1} - \zeta_{2.2}}{2} & \zeta_{1.2} & \frac{\zeta_{1.3}}{2} \\ \frac{-\zeta_{2.3}}{2} & \frac{\zeta_{1.3}}{2} & 0 \end{Vmatrix}.$$

The characteristic polynomial of this matrix coincides with the characteristic polynomial on the extremal geodesic torsion (see Section 1.12) and has the following form:

$$-\lambda^3 + \lambda^2(\tau_3, \operatorname{curl}\tau_3) - \lambda M + T = 0,$$

where M and T are the mixed and the total geodesic torsions of τ_3. As τ_3 is holonomic, then $(\tau_3, \operatorname{curl}\tau_3) = 0$. Therefore, the sum of roots of this equation is zero $\lambda_1 + \lambda_2 + \lambda_3 = 0$. Equation (9) gives $\xi_3 = 0$ at P_0. Hence, (12) at P_0 has the form of the equation on principal directions

$$-\xi_1^2\zeta_{2.1} + \xi_1\xi_2(\zeta_{1.1} - \zeta_{2.2}) + \xi_2^2\zeta_{1.2} = 0.$$

Both of the principal directions τ_1, τ_2 are the solutions of the latter equation.

Find now another form of equations of the Lamé family. Let u_i denote the derivative of u in x_i, D_u the derivative along grad u, i.e.

$$D_u = \sum_i u_i \frac{\partial}{\partial x_i}.$$

Suppose that the level surfaces of functions u, v, w generate the triorthogonal system. Differentiating the equality $\sum_i u_i v_i = 0$ in x_m, we get

$$D_u v_m + D_v u_m = 0. \tag{13}$$

In the same way we find

$$D_u w_m + D_w u_m = 0. \tag{14}$$

Differentiating the equality $\sum_m w_m v_m = 0$ along grad u, we obtain $\sum_m(w_m D_u v_m + v_m D_u u_m) = 0$. Using (13) and (14) we see that

$$\sum_m (w_m D_v u_m + v_m D_w u_m) = 0.$$

Changing the notation of the indices, we can write the last equation in the following form:

$$\sum_{m,n} v_m w_n u_{mn} = 0. \tag{15}$$

Apply to (15) the operator D_u which performs differentiation and use (13) and (14). Appropriately changing the summation indices, we obtain

$$\sum_{m,n} v_m w_n \left(D_u u_{mn} - 2\sum_k u_{mk} u_{kn} \right) = 0. \tag{16}$$

Introduce the values A_{mn} which depend on derivatives of u up to the third order:

$$A_{mn} = D_u u_{mn} - 2\sum_k u_{mk} u_{kn}.$$

Evidently, $A_{mn} = A_{nm}$. We can rewrite equation (16) briefly as:

$$\sum_{m,n} v_m w_n A_{mn} = 0. \tag{17}$$

For the following six values

$$\alpha_1 = v_1 w_1, \quad \alpha_2 = v_1 w_2 + v_2 w_1,$$
$$\alpha_3 = v_2 w_2, \quad \alpha_4 = v_1 w_3 + v_3 w_1,$$
$$\alpha_5 = v_2 w_3 + v_3 w_2, \quad \alpha_6 = v_3 w_3,$$

we have three linear homogeneous equations

$$\begin{aligned} \alpha_1 + \alpha_2 + \alpha_3 &= 0, \\ u_{11}\alpha_1 + u_{12}\alpha_2 + u_{22}\alpha_3 + u_{13}\alpha_4 + u_{23}\alpha_5 + u_{33}\alpha_6 &= 0, \\ A_{11}\alpha_1 + A_{12}\alpha_2 + A_{22}\alpha_3 + A_{13}\alpha_4 + A_{23}\alpha_5 + A_{33}\alpha_6 &= 0. \end{aligned} \tag{18}$$

Also, if we involve the equations

$$u_1 v_1 + u_2 v_2 + u_3 v_3 = 0,$$
$$u_1 w_1 + u_2 w_2 + u_3 w_3 = 0,$$

then, multiplying the first by w_1 and the second by v_1 and adding them, we obtain

$$2u_1 v_1 w_1 + u_2(v_1 w_2 + v_2 w_1) + u_3(v_1 w_3 + v_3 w_1) = 0. \tag{19}$$

In an analogous manner, multiplying by w_2 and v_2 and then by w_3 and v_3, we obtain two more equations:

$$\begin{aligned} u_1(v_1 w_2 + v_2 w_1) + 2u_2 v_2 w_2 + u_3(v_2 w_3 + v_3 w_2) &= 0, \\ u_1(v_1 w_3 + v_3 w_1) + u_2(v_2 w_3 + v_3 w_2) + 2u_3 v_3 w_3 &= 0. \end{aligned} \tag{20}$$

Thus, α_i satisfy the system of six equations (18), (19), (20). Since α_i are not equal to zero simultaneously the determinant of the system is zero:

$$\begin{vmatrix} A_{11} & A_{22} & A_{33} & A_{23} & A_{31} & A_{12} \\ u_{11} & u_{22} & u_{33} & u_{23} & u_{31} & u_{12} \\ 1 & 1 & 1 & 0 & 0 & 0 \\ 2u_1 & 0 & 0 & 0 & u_3 & u_2 \\ 0 & 2u_1 & 0 & u_3 & 0 & u_1 \\ 0 & 0 & 2u_3 & u_2 & u_1 & 0 \end{vmatrix} = 0. \tag{21}$$

So, *we obtain the equation with respect to u which contains the derivatives of u up to the third order*. From the general theory of differential equations it follows that we can take three analytic functions $u(x_1, x_2, 0)$, $u_{x_3}(x_1, x_2, 0)$ and $u_{x_3 x_3}(x_1, x_2, 0)$ in some domain of the coordinate plane $x_3 = 0$ as initial conditions. By the Cauchy–Kovalevskaya theorem, the solution of (21) is defined uniquely after that.

Thus, the arbitrariness in the construction of a triorthogonal system of surfaces in E^3 consists of three functions of two arguments.

Now we give some examples of triorthogonal systems. Laplace used the triorthogonal net of confocal surfaces of second order in studying celestial mechanics. All of these net surfaces can be represented with a single equation

$$\frac{x_1^2}{t-a_1}+\frac{x_2^2}{t-a_2}+\frac{x_3^2}{t-a_3}=1, \tag{22}$$

where a_i are distinct constants. The value of t defines the surface. For any fixed point $P(x_1, x_2, x_2)$ except the origin we can find three values of t for which (22) is satisfied. Each value of t corresponds to the surface from the family (22) which passes through P. Indeed, to find t when x_i and a_i are fixed we have the cubic equation. Suppose that $a_1 > a_2 > a_3 > 0$. Denote by $f(t)$ the left-hand side of (22). If $t = a_i$ then $f(t)$ turns into infinity. Hence, the equation $f(t) - 1 = 0$ has three real roots t_1, t_2, t_3, where

$$t_1 > a_1 > t_2 > a_2 > t_3 > a_3.$$

Thus, the point $P(x_1, x_2, x_3)$ determines the three values of the parameter t, namely t_1, t_2, t_3. Substituting one of these values into (22), we obtain the equation of the corresponding surface. If $t = t_1$ then the surface is an ellipsoid; if $t = t_2$ then the surface is a hyperboloid of one sheet; if $t = t_3$ the surface is a two-sheeted hyperboloid.

Check that the family under consideration is triorthogonal. The normal to the surface from the family $\Phi(x_1, x_2, x_3) = \text{const}$ is grad Φ. Hence the vector

$$\left\{\frac{x_1}{t_i-a_1}, \frac{x_2}{t_i-a_2}, \frac{x_3}{t_i-a_3}\right\}$$

is normal to the surface corresponding to t_i. Consider the scalar product of the normals

$$\sum_{k=1}^{3}\frac{x_k^2}{(t_i-a_k)(t_j-a_k)}=\frac{1}{t_j-t_i}\sum_{k=1}^{3}\left(\frac{x_k^2}{t_i-a_k}-\frac{x_k^2}{t_j-a_k}\right)$$
$$=\frac{1}{t_j-t_i}(f(t_i)-f(t_j))=0.$$

Therefore, the normals to the family surfaces which pass through the fixed point P are orthogonal to each other, i.e. the family is triorthogonal.

Consider the curvilinear coordinates in a domain G which can be introduced with the help of a triorthogonal family of surfaces $u_i = \text{const}$. For each point $P \in G$ we put into correspondence the numbers (u_1, u_2, u_3), namely, the values of the parameters u_i which determine three surfaces from the system each of which passes through P. We represent a position vector $\mathbf{r}$ of P as a vector function of u_i:

$$\mathbf{r} = \mathbf{r}(u_1, u_2, u_3).$$

If one of the parameters is fixed, i.e. if we set $u_i = \text{const}$, then in varying the other parameters the position vector end-point moves in the surface from a given triorthogonal system. If we fix two of the parameters, i.e. if we set $u_i = \text{const}$ and

$u_j = \text{const}$, then in varying the left parameter the point P moves along the line of intersection of those surfaces. Thus, the surfaces $u_i = \text{const}$ are the coordinate ones and their intersections are the coordinate curves. The tangent vector of the coordinate curve of parameter u_i is $\mathbf{r}_{u_i}$. Since the system is triorthogonal, $(\mathbf{r}_{u_i}, \mathbf{r}_{u_j}) = 0$ if $i \neq j$. Therefore, the first fundamental form of E^3 with respect to the curvilinear coordinates u_i has the form

$$ds^2 = (\mathbf{r}_{u_1}\, du_1 + \mathbf{r}_{u_2}\, du_2 + \mathbf{r}_{u_3}\, du_3)^2$$
$$= \mathbf{r}_{u_1}^2\, du_1^2 + \mathbf{r}_{u_2}^2\, du_2^2 + \mathbf{r}_{u_3}^2\, du_3^2.$$

Introduce the Lamé coefficients $\mathbf{r}_{u_i}^2 = H_i^2$. Then ds^2 has the form

$$ds^2 = H_1^2\, du_1 + H_2^2\, du_2^2 + H_3^2\, du_3^2.$$

The coefficients H_i are functions of u_i. As we introduce the curvilinear coordinates in a domain G of flat space, the components of the Riemann tensor R_{ijkl} are zero with respect to u_i. This gives six Lamé equations for functions H_i.

To write them in a brief form, we introduce the Darboux symbols

$$\beta_{ij} = \frac{1}{H_i}\frac{\partial H_j}{\partial u_i}, \quad i \neq j.$$

Then the Lamé equations have the form

$$\frac{\partial \beta_{ij}}{\partial u_i} + \frac{\partial \beta_{ji}}{\partial u_j} + \beta_{ki}\beta_{kj} = 0, \qquad (23)$$
$$\frac{\partial \beta_{ij}}{\partial u_k} = \beta_{ik}\beta_{kj}.$$

where i, j, k are distinct and of the range $1, 2, 3$. The first of these equations corresponds to the equality $R_{ijij} = 0$, the second one corresponds to the equality $R_{ijik} = 0$.

Let us find the expression of principal curvatures of the surface $x_3 = \text{const}$ in terms of Lame coefficients and Darboux symbols. The unit vector $\xi = \mathbf{r}_{u_3}/H_3$ is the unit normal to that surface. Therefore, the second fundamental form of the surface is

$$\mathbf{II}_3 = \sum_{i,j=1}^{2} \frac{(\mathbf{r}_{u_i u_j}, \mathbf{r}_{u_3})}{H_3}\, du_i\, du_j.$$

We denote the normal curvature of the u_1 curve in the surface $u_3 = \text{const}$ by k_{31}. This is equal to the ratio of the second and first fundamental forms, i.e.

$$k_{31} = \frac{(\mathbf{r}_{u_1 u_1}, \mathbf{r}_{u_3})}{H_3 H_1^2} = -\frac{(\mathbf{r}_{u_1}, \mathbf{r}_{u_3 u_1})}{H_3 H_1^2} = -\frac{1}{2}\frac{(\mathbf{r}_{u_1}^2)_{u_3}}{H_3 H_1^2}$$
$$= -\frac{1}{H_1 H_3}\frac{\partial H_1}{\partial u_3} = -\frac{\beta_{31}}{H_1}.$$

In an analogous manner we can find that

$$k_{ji} = -\frac{\beta_{ji}}{H_i}. \tag{24}$$

where k_{ji} means the normal curvature of the curve u_i in the surface $u_j = \text{const}$. We shall use this formula in the next section.

1.19 The Triorthogonal Bianchi System

We shall apply the Lamé equations to study the triorthogonal system, considered by Bianchi, which contains *the family of surfaces of constant negative Gaussian curvature*. Assume that the family is $u_3 = \text{const}$ and the Gaussian curvature of the family surfaces is K_0. Denote by ds_3^2 the first fundamental form of the surface $u_3 = \text{const}$:

$$ds_3^2 = H_1^2\,du_1^2 + H_2^2\,du_2^2.$$

The product of principal curvatures k_{31} and k_{32} is equal to K_0. Therefore

$$\frac{\beta_{31}\beta_{32}}{H_1 H_2} = K_0.$$

Suppose that $K_0 < 0$. Set

$$\frac{\beta_{31}}{H_1} = -\tan\omega\sqrt{-K_0}. \tag{1}$$

Then

$$\frac{\beta_{32}}{H_2} = \cot\omega\sqrt{-K_0}. \tag{2}$$

Using the Lamé equations $\frac{\partial\beta_{31}}{\partial u_2} = \beta_{32}\beta_{21}$ and $\frac{\partial\beta_{32}}{\partial u_1} = \beta_{31}\beta_{12}$, we find

$$\frac{1}{H_1}\frac{\partial H_1}{\partial u_2} = -\tan\omega\frac{\partial\omega}{\partial u_2}, \quad \frac{1}{H_2}\frac{\partial H_2}{\partial u_1} = \cot\omega\frac{\partial\omega}{\partial u_1}.$$

Solving them, we get

$$H_1 = \cos\omega\,\psi(u_1, u_3), \quad H_2 = \sin\omega\varphi(u_2, u_3).$$

where ψ does not depend on u_2 while φ does not depend on u_1.We assert that these functions do not depend on u_3. Indeed, from (1) and (2) we can find two expressions for H_3:

$$H_3 = \left[\frac{\partial\omega}{\partial u_3} - \cot\omega\frac{\partial\ln\psi}{\partial u_3}\right]\frac{1}{\sqrt{-K_0}},$$

$$H_3 = \left[\frac{\partial\omega}{\partial u_3} + \tan\omega\frac{\partial\ln\varphi}{\partial u_3}\right]\frac{1}{\sqrt{-K_0}}.$$

Hence, the following equation holds:

$$\tan\omega\,\frac{\partial\ln\varphi}{\partial u_3}+\cot\omega\,\frac{\partial\ln\psi}{\partial u_3}=0.$$

If $\frac{\partial\varphi}{\partial u_3}\neq 0$ then the latter equation implies

$$\tan\omega=\frac{M(u_1,u_3)}{N(u_2,u_3)}.$$

Consider some fixed surface $u_3=\text{const}$ and, preserving the same coordinate net, parametrize it as follows:

$$\rho_1=\int\psi(u_1,u_3)\,du_1,\quad \rho_2=\int\varphi(u_2,u_3)\,du_2.$$

Then, the first fundamental form of that surface becomes very simple

$$ds_3^2=\cos^2\omega\,d\rho_1^2+\sin^2\omega\,d\rho_2^2.$$

The angle ω is defined by the relation

$$\tan\omega=\frac{A}{B},\tag{3}$$

where A does not depend on ρ_2 while B does not depend on ρ_1. Since the curvature of $u_3=\text{const}$ is equal to K_0, the Gauss equation for this surface has the form

$$\frac{\partial^2\omega}{\partial\rho_1^2}-\frac{\partial^2\omega}{\partial\rho_2^2}=-\sin\omega\cos\omega\,K_0.\tag{4}$$

We are going to show that (3) and (4) imply the degeneration of A or B into constants. Indeed, they imply the equation

$$\left(\frac{A''}{A}+\frac{B''}{B}\right)(A^2+B^2)=-K_0(A^2+B^2)+2A'^2+2B'^2,\tag{5}$$

where the prime over A means the derivative in ρ_1 while the prime over B means the derivative in ρ_2. Differentiating (5) in ρ_1 first and then the result in ρ_2, we obtain

$$\left(\frac{A''}{A}\right)'BB'+\left(\frac{B''}{B}\right)'AA'=0.$$

The latter equation has separable variables. Thus, there is a constant C such that

$$\left(\frac{A''}{A}\right)'=CAA',\quad \left(\frac{B''}{B}\right)'=-CBB'.$$

Integrating these relations, we get

$$2A'^2=\frac{CA^4}{2}+2C_1A^2+C_3,\quad \frac{A''}{A}=\frac{CA^2}{2}+C_1,$$

$$2B'^2 = -\frac{CB^4}{2} + 2C_2B^2 + C_4, \quad \frac{B''}{B} = -\frac{CB^2}{2} + C_2,$$

where C_i are constants. Substituting these expressions into (5), we find

$$(C_2 - C_1 + K_0)A^2 + (C_1 - C_2 + K_0)B^2 = C_3 + C_4.$$

The latter equation can be satisfied if and only if either A or B is constant on the surface $u_3 = \text{const}$ and one of the coefficients of A^2 and B^2 is zero. In the latter case the surface $u_3 = \text{const}$ is a surface of revolution.

So, in general, when the surfaces of a Lamé family are not surfaces of revolution, the following equations are satisfied:

$$\frac{\partial \psi}{\partial u_3} = 0, \quad \frac{\partial \varphi}{\partial u_3} = 0.$$

Then for the Lamé coefficient H_3 we have $H_3 = \frac{\partial \omega}{\partial u_3} / \sqrt{-K_0}$. *With respect to new parametrization the first fundamental form of space can be represented as*

$$ds^2 = \cos^2\omega \, d\rho_1^2 + \sin^2\omega \, d\rho_2^2 + \left(\frac{1}{\sqrt{-K_0}} \frac{\partial \omega}{\partial \rho_3}\right)^2 d\rho_3^2.$$

So, the Lamé coefficients depend on a single function ω and the derivative $\frac{\partial \omega}{\partial \rho_3}$. The function ω satisfies the system of equations which can easily be obtained from the Lamé system. The solution of that system, as Bianchi proved (see [13]), depends on five arbitrary functions of one argument. From the expression just obtained for the first fundamental form of E^3 it follows that *the Bianchi system formed with regular surfaces with $K_0 \neq 0$ cannot exist in a domain of large size and uniform expansion in all directions.* Suppose, for instance, that the domain is a ball of radius r. Show that r is bounded from above. To do this, take a curve ρ_3 which passes through the center of the ball. Then the length of that piece of the curve which is located inside the ball which the boundary included is not less then $2r$. Since for the regular family of surfaces we have $H_3 \neq 0$, by means of the choice of positive direction in the ρ_3 curve we can fulfill the inequality $\frac{\partial \omega}{\partial \rho_3} > 0$. Let P_i be the end-points of the curve under consideration which are located in the boundary of the ball. We have

$$\frac{1}{\sqrt{-K_0}}(\omega(P_2) - \omega(P_1)) = \int \frac{1}{\sqrt{-K_0}} \frac{\partial \omega}{\partial \rho_3} \, d\rho_3 \geq 2r.$$

The angle ω satisfies the inequalities $0 < \omega < \pi/2$. Otherwise, i.e. if $\omega = 0$ or $\omega = \pi/2$ at some point, then at this point one of the first fundamental form coefficients becomes zero. But this means that in the surface $\rho_3 = \text{const}$ there is a singular point. By hypothesis the Bianchi family is formed with the regular surfaces. Hence, $r \leq \pi/4\sqrt{-K_0}$.

1.20 Geometrical Properties of the Velocity Field of an Ideal Incompressible Liquid

Vector fields occur in various problems of mathematics and mechanics. We shall consider geometrical properties of the motion of the velocity field of a liquid. The motion of a continuous medium can be described, by the Euler method, in terms of the velocity $\mathbf{V}$ of particles of the medium as a function of time t and coordinates x, y, z of points in space where the movement takes place, i.e. by the vector field

$$\mathbf{V} = \mathbf{V}(x, y, z, t).$$

The values x, y, z, t are called the *Euler variables*. The velocity field is called *stationary* if $\mathbf{V}$ does not depend on t, otherwise it is called *non-stationary*. At fixed t, the curves tangent at each point (x, y, z) to the velocity vector $\mathbf{V}(x, y, z, t)$ are called *streamlines*. The individual medium particle describes the motion of some trajectory which is tangent to $\mathbf{V}(x, y, z, t)$ at a point (x, y, z) for each t. If the velocity field is stationary then the streamlines coincide with trajectories of the motion of particles of the medium, otherwise they do not since the streamlines are different for different values of time.

For an ideal incompressible liquid the velocity field $\mathbf{V}$ *satisfies the following fundamental system of equations which were found by Euler*:

$$\operatorname{div}\mathbf{V} = 0 \qquad \text{(the equation of incompressibility)},$$

$$\frac{\partial \mathbf{V}}{\partial t} + \nabla_{\mathbf{V}}\mathbf{V} = \mathbf{F} - \frac{1}{\rho}\operatorname{grad} p \qquad \text{(the equation of tension, the Euler equation)}. \tag{1}$$

Here $\nabla_{\mathbf{V}}$ is a derivative along $\mathbf{V}$. $\mathbf{F}$ is a density of solid forces, i.e. $\mathbf{F} = \lim_{\Delta m \to 0} \frac{\Delta \mathbf{R}}{\Delta m}$, where $\Delta\mathbf{R}$ is the resultant force vector affixed to the points in a small volume $\Delta\tau$, $\Delta m = \rho\Delta\tau$ is a mass in that small volume, where ρ is the liquid density being constant for given liquid. The function p — *the hydrodynamic pressure* — depends on coordinates and time; it characterizes the action of the forces onto every infinitesimal plane area passing through the given point at the given instant.

System (1) is a system of four equations with respect to four functions, namely, three components of $\mathbf{V}$ and the p. To find a specific solution we must impose the boundary conditions and the initial conditions, in addition, for the non-stationary case. It is supposed that the solid force $\mathbf{F}$ is a given vector function of x, y, z.

Generally, the solid forces assumed the potential

$$\mathbf{F} = -\operatorname{grad} U.$$

Then the second equation in (1) can be reduced to the following form

$$\frac{\partial \mathbf{V}}{\partial t} + \operatorname{grad}\left(\frac{\mathbf{V}^2}{2} + \frac{p}{\rho} + U\right) + [\operatorname{curl}\mathbf{V}, \mathbf{V}] = 0.$$

The function $H = \frac{v^2}{2} + \frac{p}{\rho} + U$ is called the *Bernoulli function*. We shall consider a stationary velocity field. For the stationary velocity field in the motion of an ideal incompressible liquid the following system of equations holds:

$$\begin{aligned} \operatorname{div} \mathbf{V} &= 0, \\ \operatorname{grad} H &= [\mathbf{V}, \operatorname{curl} \mathbf{V}]. \end{aligned} \tag{1'}$$

From this it follows that $\mathbf{V}$ and $\operatorname{curl} \mathbf{V}$ are tangent to the surface $H = \text{const}$. Let $\mathbf{n}(x, y, z)$ be a unit vector field along $\mathbf{V}$, i.e. $\mathbf{V} = v\mathbf{n}$, where v is the value of the velocity of liquid motion. Settle the necessary geometrical properties of the field $\mathbf{n}$ in the case of a stationary velocity field. The first equation in (1) gives

$$\operatorname{div} \mathbf{V} = v \operatorname{div} \mathbf{n} + (\mathbf{n}, \operatorname{grad} v) = 0.$$

If we denote by $\frac{d}{ds}$ the derivative along the field $\mathbf{n}$ streamlines then

$$\operatorname{div} \mathbf{n} = -\frac{d \ln v}{ds}. \tag{2}$$

We see that if the velocity value is constant along the field n streamlines then the mean curvature of the field $\mathbf{n}$ is zero.

Now transform the second equation. Using the formula

$$\operatorname{curl} v\mathbf{n} = v \operatorname{curl} \mathbf{n} + [\operatorname{grad} v, \mathbf{n}].$$

we get

$$\begin{aligned} \operatorname{grad} H &= v^2[\mathbf{n}, \operatorname{curl} \mathbf{n}] + v[\mathbf{n}, [\operatorname{grad} v, \mathbf{n}]] \\ &= v^2[\mathbf{n} \operatorname{curl} \mathbf{n}] - v((\mathbf{n}, \operatorname{grad} v)\mathbf{n} - \operatorname{grad} v). \end{aligned}$$

Taking into account (2), we obtain

$$\operatorname{grad} H = v^2[\mathbf{n}, \operatorname{curl} \mathbf{n}] + v^2\mathbf{n} \operatorname{div} \mathbf{n} + v \operatorname{grad} v.$$

Since the field $\mathbf{n}$ streamline curvature vector $\mathbf{k}$ is equal to $-[\mathbf{n}, \operatorname{curl} \mathbf{n}]$, due to the Hamilton formula, which we stated in Section 1.3, we can write

$$\operatorname{grad} H = v^2(\mathbf{n} \operatorname{div} \mathbf{n} - \mathbf{k}) + \frac{1}{2} \operatorname{grad} v^2.$$

So, we have

$$\mathbf{n} \operatorname{div} \mathbf{n} - \mathbf{k} = \frac{1}{v^2} \operatorname{grad} \left(H - \frac{1}{2} v^2\right) = \frac{1}{v^2} \operatorname{grad} \left(\frac{p}{\rho} + U\right). \tag{3}$$

Relation (3) shows that the field $\mathbf{n} \operatorname{div} \mathbf{n} - \mathbf{k}$ is holonomic. The vector field $\mathbf{l} = \mathbf{n} \operatorname{div} \mathbf{n} - \mathbf{k}$ is called *the field of vectors adjoint to* $\mathbf{n}$. Observe that the field of adjoint vectors occurred in Section 1.8. It was proved there that the divergence of this field is equal to double the total curvature of the second kind, i.e. $\operatorname{div} \mathbf{l} = 2K$.

Now settle the second necessary geometrical condition for the field $\mathbf{n}$. To do this in the case $\operatorname{div}\mathbf{n} \neq 0$, we define the secondary adjoint field $\mathbf{l}^*$ as

$$\mathbf{l}^* = \mathbf{l} - \frac{[\operatorname{curl}\mathbf{l}, \mathbf{n}]}{2\operatorname{div}\mathbf{n}}.$$

Then the second necessary condition will have the form

$$\operatorname{curl}\mathbf{l}^* = 0; \tag{4}$$

more precisely, the field $\mathbf{l}^*$ is a field of gradient vectors of the function $-\ln v$:

$$\mathbf{l}^* = -\operatorname{grad}\ln v.$$

Since

$$\mathbf{l} = \frac{1}{v^2}\left(\operatorname{grad}H - \operatorname{grad}\frac{v^2}{2}\right),$$

we can write

$$\operatorname{curl}\mathbf{l} = -\frac{2}{v^3}[\operatorname{grad}v, \operatorname{grad}H].$$

Next we get

$$\frac{[\operatorname{curl}\mathbf{l}, \mathbf{n}]}{2\operatorname{div}\mathbf{n}} = -\frac{[[\operatorname{grad}v, \operatorname{grad}H], \mathbf{n}]}{v^3\operatorname{div}\mathbf{n}}$$
$$= -\frac{(\operatorname{grad}v, \mathbf{n})\operatorname{grad}H - (\operatorname{grad}H, \mathbf{n})\operatorname{grad}v}{v^3\operatorname{div}\mathbf{n}}.$$

By virtue of the second equation in (1), we have $(\operatorname{grad}H, \mathbf{n}) = 0$. Using (2), we obtain

$$\frac{[\operatorname{curl}\mathbf{l}, \mathbf{n}]}{2\operatorname{div}\mathbf{n}} = \frac{\operatorname{grad}H}{v^2}.$$

Then

$$\mathbf{l}^* = \mathbf{l} - \frac{[\operatorname{curl}\mathbf{l}, \mathbf{n}]}{2\operatorname{div}\mathbf{n}} = \frac{1}{v^2}\left(\operatorname{grad}H - \operatorname{grad}\frac{v^2}{2}\right) - \frac{\operatorname{grad}H}{v^2}$$
$$= -\operatorname{grad}\ln v. \tag{5}$$

Hence, the vector field $\mathbf{l}^*$ is a gradient and therefore $\operatorname{curl}\mathbf{l}^* = 0$.

Let P and Q be nearby points in a domain of a stream and γ be a curve joining P and Q. If $d\mathbf{r}$ is the differential of the position vector of that curve then

$$\ln v(P) = \ln v(Q) - \int_{\gamma}(\mathbf{l}^*, d\mathbf{r}).$$

Thus, the function $\ln v$ is defined by its value at one point and the geometry of flow, i.e. by the geometry of the vector field $\mathbf{n}$.

Condition (2) follows from equation (5). Indeed, $(\mathbf{I}^*, \mathbf{n}) = (\mathbf{I}, \mathbf{n}) = \operatorname{div}\mathbf{n}$.

Conditions (3) and (4) were stated by S. Bushgens. So, the following theorem holds.

Theorem *If* $\mathbf{n}$ *is a unit vector field of the velocity directions of an ideal incompressible liquid in the field of potential forces then the vector field*

$$\mathbf{I} = \mathbf{n}\operatorname{div}\mathbf{n} - \mathbf{k}$$

is holonomic, and the vector field

$$\mathbf{I}^* = \mathbf{I} - \frac{[\operatorname{curl}\mathbf{I}, \mathbf{n}]}{2\operatorname{div}\mathbf{n}}$$

is a gradient.

Bushgens assumed that these conditions are also sufficient. But that is not true. Besides these necessary conditions, we need one more condition which is generated by the relation between functions expressing the holonomicity of $\mathbf{I}$ and the possibility of $\mathbf{I}^*$ to be expressed as a gradient. Let

$$\mathbf{I} = \lambda \operatorname{grad}\varphi, \quad \mathbf{I}^* = -\operatorname{grad}\psi.$$

The functions λ, φ and ψ are not arbitrary. They are connected with each other due to (3) and (4). From (5) we see that $v = e^{\psi}$. If $\mathbf{n}$ is a vector field of liquid flow velocity then by (3) we get

$$\lambda \operatorname{grad}\varphi = e^{-2\psi} \operatorname{grad}\left(\frac{p}{\rho} + U\right).$$

Hence, the vector field $e^{2\psi}\lambda\operatorname{grad}\varphi$ is a gradient. From this it follows that $[\operatorname{grad}\lambda e^{2\psi}, \operatorname{grad}\varphi] = 0$, i.e. $\lambda e^{2\psi}$ is some function of φ

$$\lambda e^{2\psi} = f(\varphi). \tag{6}$$

Conditions (3), (4) and (6) are necessary and sufficient for the vector field $\mathbf{n}$ *to be the field of liquid flow velocity.* Indeed, if (6) holds then

$$\lambda \operatorname{grad}\varphi = \frac{f(\varphi)\operatorname{grad}\varphi}{e^{2\psi}} = \frac{\operatorname{grad}\theta(\varphi)}{e^{2\psi}},$$

where $\theta(\varphi)$ is some function. Set $\frac{p}{\rho} + U = \theta(\varphi)$. We define the value of flow velocity by (2). Then (1′) will be satisfied.

One of the central problems in liquid theory is the problem of the existence of flow with a given boundary and initial conditions. Note that for three-dimensional flows of an ideal incompressible liquid the solution of system (1) which is global in time is not known. Therefore, for the detailed study of liquids only some classes of fields are acceptable.

One of the important classes of streams widely used in applications is a stream without rotations, i.e. streams for which

$$\operatorname{curl}\mathbf{V} = 0.$$

Such an assumption is justified by the following facts:

(1) The *Lagrange theorem* holds: *if at some initial instant we have* $\operatorname{curl}\mathbf{V} = 0$ *at all points of an ideal liquid moving under the action of solid forces with a potential then* $\operatorname{curl}\mathbf{V} = 0$ *at any subsequent instant.*

(2) In describing the motion of the body within a liquid it is supposed that the liquid is initially immovable. Since the immovable liquid contains no rotations, the motion remains irrotational.

For the irrotational motion the field $\mathbf{n}$ is holonomic. Indeed, $\operatorname{curl}\mathbf{V} = 0$ means that at each instant

$$\mathbf{V} = \operatorname{grad}\varphi,$$

i.e. the field $\mathbf{n}$ is orthogonal to the family of surfaces $\varphi = \text{const}$. The function φ depends on x, y, z and t; also, (1) implies that φ satisfies the Laplace equation

$$\varphi_{xx} + \varphi_{yy} + \varphi_{zz} = 0.$$

This equation is solvable in a given boundary condition. Next, from (1) we find

$$\operatorname{grad}\left(\frac{\partial\varphi}{\partial t} + H\right) = 0.$$

Hence

$$\frac{\partial\varphi}{\partial t} + \frac{\varphi_x^2 + \varphi_y^2 + \varphi_z^2}{2} + \frac{p}{\rho} + U = f(t), \tag{7}$$

where $f(t)$ is some function of time but independent of space coordinates. Then from (7) we find the pressure

$$p = \rho\left\{f(t) - \frac{\partial\varphi}{\partial t} - \frac{1}{2}|\operatorname{grad}\varphi|^2\right\}.$$

Another class, considered by Gromeka and Beltrami, forms the *screw streams*, i.e. the streams satisfying

$$\operatorname{curl}\mathbf{V} = \lambda\mathbf{V},$$

where λ is a scalar function. This class is also interesting from the practical viewpoint. Those streams arise in descriptions of liquid motion generated by the water-screw of a ship, in the theory of wings, propellers etc.

A necessary and sufficient condition for the vector field $\mathbf{n}$ to have a stationary screw flow is the following requirement: the adjoint field $\mathbf{I}$ is a gradient.

Indeed, for the screw flow we have an equation $\operatorname{curl}\mathbf{V} = \lambda\mathbf{V}$ or

$$v\operatorname{curl}\mathbf{n} + [\operatorname{grad} v, \mathbf{n}] = \lambda v\,\mathbf{n}.$$

Find the vector product of both sides of this equation with $\mathbf{n}$:

$$v[\operatorname{curl}\mathbf{n}, \mathbf{n}] + (\operatorname{grad} v, \mathbf{n})\mathbf{n} - \operatorname{grad} v = 0.$$

Earlier we obtained (grad $\ln v, \mathbf{n}) = -\operatorname{div}\mathbf{n}$. Hence

$$-\mathbf{I} = [\operatorname{curl}\mathbf{n}, \mathbf{n}] - \mathbf{n}\operatorname{div}\mathbf{n} = \operatorname{grad}\ln v.$$

Conversely, if $[\operatorname{curl}\mathbf{n}, \mathbf{n}] - \mathbf{n}\operatorname{div}\mathbf{n} = \operatorname{grad}\psi$, where ψ is some function, then $(\operatorname{grad}\psi, \mathbf{n}) = -\operatorname{div}\mathbf{n}$. Set the velocity of flow to be $v = e^{\psi}$. We have

$$\operatorname{div}\mathbf{V} = e^{\psi}((\operatorname{grad}\psi, \mathbf{n}) + \operatorname{div}\mathbf{n}) = 0.$$

Next,

$$\operatorname{curl}\mathbf{V} = e^{\psi}((\operatorname{grad}\psi, \mathbf{n}) + \operatorname{curl}\mathbf{n}) = (\mathbf{n}, \operatorname{curl}\mathbf{n})\mathbf{V}.$$

This means that $\mathbf{V}$ is a screw flow. Both equations for liquid flow are fulfilled and the function $H \equiv \text{const}$.

Note the following interesting property of screw flow: the non-holonomicity $(\mathbf{n}\operatorname{curl}\mathbf{n})$ is constant along the streamlines of the field $\mathbf{n}$.

1.21 The Carathéodory–Rashevski Theorem

In the case of $(\mathbf{n}, \operatorname{curl}\mathbf{n}) \equiv 0$ the curves orthogonal to the field and passing through the fixed point lie in the same surface, i.e. in this case the space is a foliated one. The set of points which can be joined to a fixed point with the curves orthogonal to the field forms a surface. In the case of a non-holonomic field we have essentially another picture. The following theorem holds.

Theorem *Let us be given a regular unit vector field* $\mathbf{n}$ *in a domain* $G \subset E^3$ *with the non-holonomicity value* $(\mathbf{n}, \operatorname{curl}\mathbf{n}) \neq 0$. *Then any two points in G can be joined by the curve orthogonal to the field* $\mathbf{n}$.

A more general statement was proved in 1909 by C. Carathéodory in thermodynamics [12]. He proved the following.

Let us be given the Pfaff equation $dx_0 + X_1\,dx_1 + \cdots + X_n\,dx_n = 0$, *where* X_i *are continuously differentiable functions of* x_j. *If in each neighborhood of an arbitrary point P in the space of* x_i *there is a point inaccessible by the curve satisfying the equation then there is a multiplier which turns the equation into a total differential.*

The proof is short but contains some vague arguments. A more general theorem was proved by P. Rashevski [29]. Let us be given m vector fields $X_1, \ldots, X_m$ in a domain G of Euclidean space E^n. Let us take all various Poisson brackets $[X_i, X_j] = \nabla_{X_j}X_i - \nabla_{X_i}X_j$.

Theorem (Rashevski) *Suppose that among the vector fields* $X_1, \ldots, X_m$ *and all their consecutive Poisson brackets it is possible to select n fields which are linearly independent at each point of G. Then it is possible to pass from every point of G to any other point of G by a finite number of moves along trajectories of the vector fields* $X_1, \ldots, X_m$.

We use the Rashevski proof and, for the sake of simplicity, consider the vector fields in E^3.

At first, we state an elementary lemma from calculus. Let (x_1, x_2, x_3) be a point in G.

Lemma 1 Let us be given a function $F(x)$ of class C^k. Suppose that x_0, x_1 are the points which can be joined by the intercept in G. Then

$$F(x) - F(x_0) = \sum_{i=1}^{3}(x_i - x_{i0})A_i,$$

where A_i are the functions of x_i and x_{i0} of class C^{k-1}.

Introduce the function of the parameter t:

$$\Phi(t) = F(x_0 + t(x - x_0)).$$

Its derivative is

$$\frac{d\Phi}{dt} = \sum_{i=1}^{3} F_{x_i}(x_i - x_{i0}).$$

Then

$$F(x) - F(x_0) = \int_0^1 \frac{d\Phi}{dt}\,dt$$

$$= \sum_{i=1}^{3}(x_i - x_{i0}) \int_0^1 F_{x_i}(x_0 + t(x - x_0))\,dt.$$

Denote the integral coefficients of $(x_i - x_{i0})$ by A_1, A_2, A_3. Note that $A_i = F_{x_i}(x_0)$ for $x_i = x_{i0}$. The lemma is proved.

Suppose that there are given in G the vector fields depending on the parameters t_1 and t_2, namely

$$\mathbf{l}(x, t_1) \quad \text{and} \quad \mathbf{m}(x, t_2).$$

We can regard the field $\mathbf{l}(x, t)$ as a mapping which maps x into the $\mathbf{l}$ end-point. We suppose that the following conditions are fulfilled:

(I) There are continuous derivatives of those fields up to the k-th order with respect to each argument.

(II) There is an inverse vector field. The inverse vector field $\mathbf{l}^{-1}(x, t_1)$ is that with the following property: if $t_1\mathbf{l}(x, t)$ maps x into x' then $t_1\mathbf{l}^{-1}(x', t)$ maps $x' \to x$. Denote by $\mathbf{l}_0$ the field $\mathbf{l}(x, 0)$. By means of the fields $\mathbf{l}$ and $\mathbf{m}$ we are able to construct one more vector field ∇ as follows: shift the point x into y by $t_1\mathbf{l}^{-1}(x, t_1)$, then shift y into z by $t_2\mathbf{m}^{-1}(y, t_2)$, next shift z into u by $t_1\mathbf{l}(z, t_1)$, finally shift u into v by $t_2\mathbf{m}(u, t_2)$. The difference between v and x is not equal to zero. As a matter of fact, this difference depends on x, t_1 and t_2. Denote it by $\nabla(x, t_1, t_2)$:

$$\nabla(x, t_1, t_2) = t_1(\mathbf{l}(z, t_1) - \mathbf{l}(y, t_1)) + t_2(\mathbf{m}(u, t_2) - \mathbf{m}(z, t_2)).$$

By the lemma, we can write

$$\mathbf{l}(z, t_1) - \mathbf{l}(y, t_1) = \sum_{i=1}^{3}(z_i - y_i)\mathbf{L}_3,$$

where the $\mathbf{L}_i$ are the vector functions of y, z. Also, if z coincides with y then L_i coincides with $\frac{\partial \mathbf{l}(y, t_1)}{\partial y_i}$. Analogously

$$\mathbf{m}(u, t_2) - \mathbf{m}(z, t_2) = \sum_{i=1}^{3}(u_i - z_i)M_i.$$

If $u = z$ then

$$M_i = \frac{\partial \mathbf{m}(z, t_2)}{\partial z_i}.$$

The components $\{z_i - y_i\}$ are the components of the shift from y to z, i.e. of $-t_2\mathbf{m}(z, t_2)$. Let $m_i(z, t_2)$ be the components of $\mathbf{m}(z, t_2)$. Analogously, the components $\{u_i - z_i\}$ are the components of the shift from z to u, i.e. of the vector $t_1\mathbf{l}(z, t_1)$. Set $\mathbf{l}(z, t_1) = \{l_i(z, t_1)\}$. So, ∇ can be expressed as

$$\begin{aligned}\nabla &= -t_1t_2\mathbf{m}_i(z, t_2)L_i + t_1t_2\mathbf{l}_i(z, t_1)M_i\\ &= t_1t_2\{\mathbf{l}_i(z, t_1)M_i - \mathbf{m}_i(z, t_2)L_i\}.\end{aligned}$$

We denote the expression in braces by $\eta(x, t_1, t_2)$. Note that η is constructed by means of $\mathbf{l}$ and $\mathbf{m}$. The field $\eta(x, t_1, t_2)$ is invertible. To find the $\eta^{-1}(x, t_1, t_2)$ we shall make the chain of shifts: $x \to y'$ by $t_2\mathbf{m}^{-1}(x, t_2)$, $y' \to z'$ by $t_1\mathbf{l}^{-1}(y', t_1)$, $z' \to u'$ by $t_2\mathbf{m}(z', t_2)$, $u' \to v'$ by $t_1\mathbf{l}(u', t_1)$. Now find the η_0, i.e. the function $\eta(x, t_1, t_2)$ for

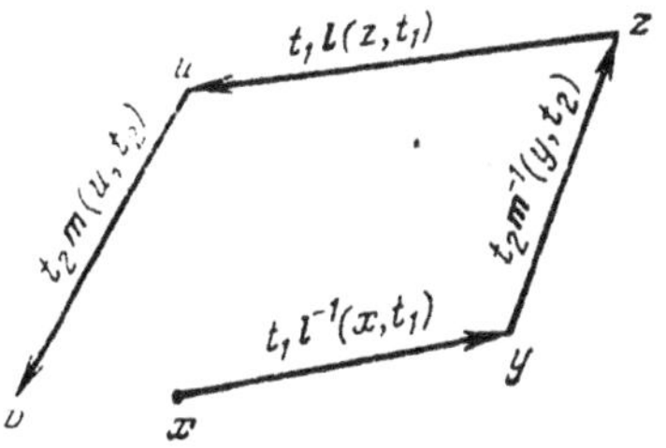

FIGURE 20

$t_1 = t_2 = 0$. We have $\mathbf{l}(z, 0) = \mathbf{l}_0$. But $z(x) = x$ for $t_1 = 0$. Also M_i turns into a partial derivative of $\mathbf{m}$: $M_{i0} = \frac{\partial \mathbf{m}(x,0)}{\partial x_i}$. Analogously, $m_i(z, 0) = m_{i0}$, $L_{i0} = \frac{\partial \mathbf{l}(x,0)}{\partial x_i}$. Hence

$$\eta_0(x) = \mathbf{l}_{i0} M_{i0} - \mathbf{m}_{i0} L_{i0} = \mathbf{l}_{i0}(x) \frac{\partial \mathbf{m}(x,0)}{\partial x_i} - \mathbf{m}_{i0}(x) \frac{\partial \mathbf{l}(x,0)}{\partial x_i}.$$

If we interpret the vector fields $\mathbf{l}_0, \mathbf{m}_0, \eta_0$ as the fields of operators

$$\mathbf{l}_0 = \sum_i \mathbf{l}_{i0} \frac{\partial}{\partial x_i},$$

then $\eta_0(x)$ can be interpreted as the Poisson bracket of operators $\mathbf{l}_0$ and $\mathbf{m}_0$.

We say that a point y is accessible from x by means of some set of vector functions $\mathbf{l}(x, t_1), \mathbf{m}(x, t_2), \ldots$ if we are able to pass from x to y as a result of shifts of the following kind: a shift from x into x' by $t_1 \mathbf{l}(x, t_1)$, then a shift from x' into x'' by $t_2 \mathbf{m}(x', t_2)$ and so on; or use the vectors $t_1 \mathbf{l}^{-1}(x, t_1)$, $t_2 \mathbf{m}^{-1}(x, t_2)$ in the procedure above. Since the vector field $\eta(x, t_1, t_2)$ has been constructed by means of $\mathbf{l}$ and $\mathbf{m}$, then the point which is accessible by means of $\mathbf{l}$, $\mathbf{m}$ and η will be accessible by means of $\mathbf{l}$ and $\mathbf{m}$.

Lemma 2 Let us be given the vector functions $\mathbf{l}(x, t_1)$, $\mathbf{m}(x, t_2)$, $\eta(x, t_1, t_2)$ in $G \subset E^3$. Suppose that the corresponding $\mathbf{l}_0$, $\mathbf{m}_0$, η_0 are linearly independent at each point of G. Then any two points are mutually accessible by means of $\mathbf{l}, \mathbf{m}, \eta$.

We shall prove that any point in some neighborhood of x is accessible. Let x, y, z, v be a sequence of points formed as follows:

y is a result of a shift from x by $t_1 \mathbf{l}(x, t_1)$,
z is a result of a shift from y by $t_2 \mathbf{m}(y, t_2)$,
v is a result of a shift from z by $\tau_1 \tau_2 \eta(z, \tau_1, \tau_2)$.

We obtain

$$\mathbf{v}(t_1, t_2, \tau_1, \tau_2) = \mathbf{x} + t_1 \mathbf{l}(x, t_1) + t_2 \mathbf{m}(y, t_2) + \tau_1 \tau_2 \eta(z, \tau_1, \tau_2). \tag{1}$$

Fix $\tau_2 \neq 0$ and find $\partial \mathbf{v}/\partial t_1$ for $t_1 = t_2 = \tau_1 = 0$. Evidently,

$$\frac{\partial \mathbf{v}}{\partial t_1}(0, 0, 0, \tau_2) = \mathbf{l}(x, 0) = \mathbf{l}_0(x).$$

Analogously

$$\frac{\partial \mathbf{v}}{\partial t_2}(0, 0, 0, \tau_2) = \mathbf{m}(x, 0) = \mathbf{m}_0(x), \qquad \frac{\partial \mathbf{v}}{\partial \tau_1} = \tau_2 \eta_0,$$

and we see that the Jacobian of transformation from (t_1, t_2, τ_1) to (v_1, v_2, v_3) is equal to

$$\tau_2(\mathbf{l}_0, \mathbf{m}_0, \eta_0) \neq 0. \tag{2}$$

Therefore, for any point v we can find t_1, t_2, τ_1 such that (1) will be satisfied. This implies that for some sufficiently small neighborhood of x for the values of v and of 0 for the values of t_1, t_2, τ_1 there exists a one-to-one correspondence between the neighborhoods.

Let us be given in G the non-holonomic vector field $\mathbf{n}$, i.e. $(\mathbf{n}, \operatorname{curl} \mathbf{n}) \neq 0$. Take the vector fields $\mathbf{X}_1$ and $\mathbf{X}_2$ orthogonal to $\mathbf{n}$. Each of the vector fields $\mathbf{X}_1, \mathbf{X}_2$ generates the streamlines

$$\frac{d\mathbf{x}}{dt_1} = \mathbf{X}_1(x), \quad \frac{d\mathbf{x}}{dt_2} = \mathbf{X}_2(x). \tag{3}$$

Now introduce the parametric vector fields $\mathbf{l}_1(x, t_1)$ and $\mathbf{l}_2(x, t_2)$ as follows.

Draw the integral curve of (3) through each point x in G, i.e. the trajectory of the operator $\mathbf{X}_i$, where t_i is the parameter of that trajectory. We assume that $t_i = 0$ at the initial point x. Now define the shift vector $\bar{\Delta}(x, t_i)$ as a vector from x to the point in the trajectory which corresponds to the parameter value t_i. Restricting the values of t_i to some sufficiently small interval about zero, we are able to achieve the situation when for any x in a domain $G^* \subset G$ the corresponding trajectory segment exists and does not go out of the main domain G. So, $\bar{\Delta}(x, t_i)$ is already defined. Evidently,

$$\bar{\Delta}(x, t_i) = \int_0^{t_i} \mathbf{X}_i(u_1, u_2, u_3)\, d\tau_i,$$

where u_1, u_2, u_3 are the coordinates in the i-th trajectory being functions of τ_i and initial coordinates x_1, x_2, x_3. The parameter τ_i varies from 0 to t_i. Introduce a new parameter $t = \tau_i / t_i$, so $\tau_i = t t_i$. Then

$$\bar{\Delta}(x, t_i) = t_i \int_0^1 \mathbf{X}_i(u_1, u_2, u_3)\, dt,$$

where u_1, u_2, u_3 are continuously differentiable up to the k-th order functions of the initial coordinates x_1, x_2, x_3 and t, t_i. Denote the integral on the right of the expression above by $\mathbf{l}_i(x, t_i)$. Then

$$\bar{\Delta}(x, t_i) = t_i \mathbf{l}_i(x, t_i).$$

Thus, for each operator $\mathbf{X}_i$ we introduced the corresponding vector function $\mathbf{l}_i(x, t_i)$. This function is invertible: $\mathbf{l}_i^{-1}(x, t_i) = -\mathbf{l}_i(x, -t_i)$.

If $t_i = 0$ then $\mathbf{l}_{i0} = \mathbf{X}_i(x)$. To prove the theorem it is now sufficient to show that we are able to reach any point y starting from x by means of $\mathbf{l}_1(x, t_1)$ and $\mathbf{l}_2(x, t_2)$. Since $(\mathbf{n}, \operatorname{curl} \mathbf{n}) \neq 0$, η, being constructed by means of $\mathbf{l}_{10}$ and $\mathbf{l}_{20}$, is linearly independent of $\mathbf{l}_{10}$ and $\mathbf{l}_{20}$. Also, $\eta_0(x, 0, 0)$ is a Poisson bracket of the vector fields $\mathbf{l}_{10} = \mathbf{X}_1$ and $\mathbf{l}_{10} = \mathbf{X}_2$. By the previous lemma, any point is accessible by means of $\mathbf{l}_1, \mathbf{l}_2$ and η. But accessibility by means of $\mathbf{l}_1, \mathbf{l}_2, \eta$ implies accessibility by means of $\mathbf{l}_1, \mathbf{l}_2$. The theorem is proved.

1.22 Parallel Transport on the Non-Holonomic Manifold and the Vagner Vect

We shall call the set of curves orthogonal to the field **n** the *non-holonomic ma*
We say that *a vector field is tangent to the non-holonomic manifold if it is orth*
to **n**. V. Vagner introduced the notion of parallel transport of tangent vector
respect to the field **n**, i.e. parallel transport on the non-holonomic manifold [

Let us be given a vector field **v** orthogonal to **n** at each point. Consid
infinitesimal motion along the curve orthogonal to **n**. We call those curv
missible. As usual, we denote the field differential in E^3 by $d\mathbf{v}$. Denote by
projection of $d\mathbf{v}$ into the plane orthogonal to **n**:

$$\delta\mathbf{v} = d\mathbf{v} - \mathbf{n}(\mathbf{n}, d\mathbf{v}).$$

We call $\delta\mathbf{v}$ *the absolute differential of* **v** *on the non-holonomic manifold.*
$(\mathbf{n}, \mathbf{v}) = 0$ then we can rewrite (1) as

$$\delta\mathbf{v} = d\mathbf{v} + \mathbf{n}(\mathbf{v}, d\mathbf{n}).$$

The tangent vector on the non-holonomic manifold is called *parallel-transpo*
along the admissible curve if its absolute differential along the curve is zero. Fo
field

$$d\mathbf{v} = -\mathbf{n}(\mathbf{v}, d\mathbf{n}),$$

where the differentials are assumed to be along the curve under consideratior

Let us answer the question of what is the vector which is tangent to the admi
curve γ and parallel-transportable along it. If $\mathbf{r} = \mathbf{r}(s)$ is the curve γ represent
with respect to the arc length parameter s and $\mathbf{v} = \mathbf{r}_s$ is parallel-transportable,

$$\mathbf{r}_{ss} = -\mathbf{n}(\mathbf{r}_s, \mathbf{n}_s).$$

According to Section 1.4, this is the equation of the straightest lines. Conversel
is a straightest line then its tangent vector is parallel-transportable.

It is natural to ask the question: *in which case are we able to construct the*
field defined over the whole domain of definition of the non-holonomic manifol
that it will be parallel-transportable along every admissible curve? Evidently, al
single admissible curve such a field exists: this is a solution of an ordinary differ
equation with the right-hand side linear in **v**. However, the two points in space c
joined to each other with the different admissible curves and the transport along
of them gives, maybe, different results at the same point. Therefore, our que
may be reformulated as follows: *in which case does not the parallel transport d*
on the curve of the transport?

Denote by $d\mathbf{n}/dr$ and $d\mathbf{v}/ds$ the operators generated by derivatives of these f
The corresponding matrices have the form

$$\left\|\begin{matrix} \xi_1^1 & \xi_2^1 & \xi_3^1 \\ \xi_1^2 & \xi_2^2 & \xi_3^2 \\ \xi_1^3 & \xi_2^3 & \xi_3^3 \end{matrix}\right\|, \qquad \left\|\begin{matrix} v_1^1 & v_2^1 & v_3^1 \\ v_1^2 & v_2^2 & v_3^2 \\ v_1^3 & v_2^3 & v_3^3 \end{matrix}\right\|,$$

where $\xi^i_j = \frac{\partial \xi^i}{\partial x^j}$, $v^i_j = \frac{\partial v^i}{\partial x^j}$. The differentials $d\mathbf{n}$ and $d\mathbf{v}$ can be represented in terms of those operators as

$$d\mathbf{n} = \frac{d\mathbf{n}}{dr}d\mathbf{r}, \quad d\mathbf{v} = \frac{d\mathbf{v}}{dr}d\mathbf{r}.$$

Note that since $\mathbf{n}$ is a unit vector, $\frac{d\mathbf{n}}{dr}\mathbf{x}$ is always orthogonal $\mathbf{n}$, where $\mathbf{x}$ is an arbitrary vector. Let $\mathbf{a}$ be any vector field tangent to the non-holonomic manifold. For any parallel-transportable vector field $\mathbf{v}$ along any admissible curve the following equation holds:

$$\frac{d\mathbf{v}}{dr}\mathbf{a} = -\mathbf{n}\left(\mathbf{v}, \frac{d\mathbf{n}}{dr}\mathbf{a}\right). \tag{3}$$

Let $\mathbf{x}$ and $\mathbf{y}$ be constant vector fields in E^3. Denote by $\mathbf{a}$ and $\mathbf{b}$ the projections of $\mathbf{x}$ and $\mathbf{y}$ respectively into the plane orthogonal to $\mathbf{n}$. We want to extend the concept of parallel transport to arbitrary curves. To do so, find the result of the action of the operator $d\mathbf{v}/dr$ over $\mathbf{n}$. Set $A = d\mathbf{n}/dr$.

Lemma Parallel transport along admissible curves induces the parallel transport along the field $\mathbf{n}$ streamlines in the case of $(\mathbf{n}, \operatorname{curl}\mathbf{n}) \neq 0$ and this transport is defined by the following equation:

$$\frac{d\mathbf{v}}{dr}\mathbf{n} = -\mathbf{n}(\mathbf{v}, A\mathbf{n}) - \frac{K[\mathbf{n}, \mathbf{v}]}{(\mathbf{n}, \operatorname{curl}\mathbf{n})}.$$

Differentiating both sides of (3) in the direction of the vector $\mathbf{b}$, we obtain

$$\begin{aligned}\frac{d^2\mathbf{v}}{dr^2}(\mathbf{a}, \mathbf{b}) + \frac{d\mathbf{v}}{dr}\frac{d\mathbf{a}}{dr}\mathbf{b} = & - A\mathbf{b}(\mathbf{v}, A\mathbf{a}) - \mathbf{n}\left(\frac{d\mathbf{v}}{dr}\mathbf{b}, A\mathbf{a}\right) \\ & - \mathbf{n}\left(\mathbf{v}, \frac{d^2\mathbf{n}}{dr^2}(\mathbf{a}, \mathbf{b})\right) - \mathbf{n}\left(\mathbf{v}, A\frac{d\mathbf{a}}{dr}\mathbf{b}\right),\end{aligned} \tag{4}$$

where $\frac{d^2\mathbf{v}}{dr^2}$ and $\frac{d^2\mathbf{n}}{dr^2}$ are the linear vector functions of two arguments. They are formed from the second derivatives of components of the vector fields $\mathbf{v}$ and $\mathbf{n}$ respectively. For instance, the α-th component of $\frac{d^2\mathbf{v}}{dr^2}$ has the form

$$\frac{\partial^2 v^\alpha}{\partial x^i \partial x^j} a^i b^j,$$

where a^i, b^j are components of $\mathbf{a}$ and $\mathbf{b}$ respectively. The equality of mixed derivatives of v^α implies that the vector function $\frac{d^2\mathbf{v}}{dr^2}$ is symmetric in the arguments $\mathbf{a}$ and $\mathbf{b}$.

Since $\mathbf{v}$ is parallel-transportable, then $\frac{d\mathbf{v}}{dr}\mathbf{b}$ is collinear to $\mathbf{n}$, $A\mathbf{a}$ is orthogonal to $\mathbf{n}$. Therefore, the second term on the right-hand side of (4) is zero. Transfer the second term on the left of (4) into the right-hand side. Since $\mathbf{a} = \mathbf{x} - \mathbf{n}(\mathbf{n}, \mathbf{x})$ and $\mathbf{x}$ is a constant vector field,

$$\frac{d\mathbf{a}}{dr}\mathbf{b} = -A\mathbf{b}(\mathbf{n}, \mathbf{x}) - \mathbf{n}(\mathbf{x}, A\mathbf{b}). \tag{5}$$

Using (5), transform the next two terms in the right-hand side of (4):

$$-\frac{d\mathbf{v}}{dr}\frac{d\mathbf{a}}{dr}\mathbf{b} - \mathbf{n}\left(\mathbf{v}, A\frac{d\mathbf{a}}{dr}\mathbf{b}\right) = (\mathbf{n}, \mathbf{x})\left(\frac{d\mathbf{v}}{dr}A\mathbf{b} + \mathbf{n}(\mathbf{v}, A^2\mathbf{b})\right) + (\mathbf{x}, A\mathbf{b})\frac{d\mathbf{v}}{dr}\mathbf{n} + \mathbf{n}(\mathbf{x}, A\mathbf{b})(\mathbf{v}, A\mathbf{n}). \qquad (6)$$

Since $A\mathbf{b}$ is orthogonal to $\mathbf{n}$, by virtue of (3) when $\mathbf{a}$ is replaced with $A\mathbf{b}$ we obtain

$$\frac{d\mathbf{v}}{dr}A\mathbf{b} = -\mathbf{n}(\mathbf{v}, A^2\mathbf{b}).$$

Therefore, on the right-hand side of (6) the multiplier of $(\mathbf{n}, \mathbf{x})$ is zero. Taking into account all the transformations, rewrite (4) in the following form:

$$\frac{d^2\mathbf{v}}{dr^2}(\mathbf{a}, \mathbf{b}) = (\mathbf{x}, A\mathbf{b})\frac{d\mathbf{v}}{dr}\mathbf{n} + \mathbf{n}(\mathbf{x}, A\mathbf{b})(\mathbf{v}, A\mathbf{n}) - A\mathbf{b}(\mathbf{v}, A\mathbf{a}) - \mathbf{n}\left(\mathbf{v}, \frac{d^2\mathbf{n}}{dr^2}(\mathbf{a}, \mathbf{b})\right). \qquad (7)$$

Since $A\mathbf{b}$ is orthogonal to $\mathbf{n}$, $\mathbf{x}$ may be replaced with $\mathbf{a}$. Now interchange the roles of $\mathbf{a}$ and $\mathbf{b}$. Subtract the equation obtained from the original. Using the symmetry of the vector functions $d^2\mathbf{v}/dr^2$ and $d^2\mathbf{n}/dr^2$ with respect to the arguments, we find

$$\{(\mathbf{a}, A\mathbf{b}) - (\mathbf{b}, A\mathbf{a})\}\left\{\frac{d\mathbf{v}}{dr}\mathbf{n} + \mathbf{n}(\mathbf{v}, A\mathbf{n})\right\} + A\mathbf{a}(\mathbf{v}, A\mathbf{b}) - A\mathbf{b}(\mathbf{v}, A\mathbf{a}) = 0. \qquad (8)$$

Transform the expression in the first braces:

$$\begin{aligned}(\mathbf{a}, A\mathbf{b}) - (\mathbf{b}, A\mathbf{a}) &= \xi^i_j a^i b^j - \xi^i_j a^j b^i = \frac{1}{2}(a^i b^j - a^j b^i)(\xi^i_j - \xi^j_i) \\ &= -(\mathbf{a}, \mathbf{b}, \operatorname{curl}\mathbf{n}) \\ &= -(\mathbf{n}, \operatorname{curl}\mathbf{n})(\mathbf{n}, \mathbf{a}, \mathbf{b}). \qquad (9)\end{aligned}$$

We can represent the difference of the two last terms in (8) as the following vector product $[[A\mathbf{b}, A\mathbf{a}], \mathbf{v}]$. But $A\mathbf{b}$ and $A\mathbf{a}$ are orthogonal to $\mathbf{n}$. Therefore

$$[[A\mathbf{b}, A\mathbf{a}], \mathbf{v}] = -[\mathbf{n}, \mathbf{v}](A\mathbf{a}, A\mathbf{b}, \mathbf{n}).$$

Now show that the total curvature K of the field $\mathbf{n}$ (remember that K is the second symmetric function of principal curvatures of the second kind) can be found by the following formula:

$$K = \frac{(A\mathbf{a}, A\mathbf{b}, \mathbf{n})}{(\mathbf{n}, \mathbf{a}, \mathbf{b})}, \qquad (10)$$

where $\mathbf{a}$, $\mathbf{b}$ are the vectors orthogonal to $\mathbf{n}$. Since the right-hand side of (10) is invariant with respect to rotation of coordinate axes in a space, choose the coordinate axes in a special way, namely direct $\mathbf{e}_3$ along $\mathbf{n}(M)$, where M is some point. At this point $\xi^3 = 1$, $\xi_i^3 = 0$, $a^3 = b^3 = 0$. We have

$$(A\mathbf{a}, A\mathbf{b}, \mathbf{n}) = \begin{vmatrix} \xi_i^1 a^i & \xi_i^2 a^i & 0 \\ \xi_j^1 b^j & \xi_j^2 b^j & 0 \\ 0 & 0 & 1 \end{vmatrix} = (\xi_1^1\xi_2^2 - \xi_2^1\xi_1^2)(a^1b^2 - b^1a^2)$$
$$= K(\mathbf{n}, \mathbf{a}, \mathbf{b}),$$

which completes the proof of (10). On account of (10) we get

$$[[A\mathbf{b}, A\mathbf{a}], \mathbf{v}] = -K[\mathbf{n}, \mathbf{v}](\mathbf{n}, \mathbf{a}, \mathbf{b}). \tag{11}$$

By the hypothesis of the lemma, $(\mathbf{n}, \operatorname{curl}\mathbf{n}) \neq 0$. Taking into account (9) and (11), the equation (8) can be represented as

$$\frac{d\mathbf{v}}{dr}\mathbf{n} = -\mathbf{n}(\mathbf{v}, A\mathbf{n}) - \frac{K[\mathbf{n}, \mathbf{v}]}{(\mathbf{n}, \operatorname{curl}\mathbf{n})}. \tag{12}$$

Thus, we obtain the value of the action of the operator $d\mathbf{v}/dr$ over $\mathbf{n}$. The Lemma is proved.

Decomposing any vector $\mathbf{x}$ into components collinear and orthogonal to $\mathbf{n}$, we are able to give the equation of parallel transport with respect to each of them. Adding those equations, we get

$$\frac{d\mathbf{v}}{dr}\mathbf{x} = -\mathbf{n}(\mathbf{v}, A\mathbf{x}) - \frac{K[\mathbf{n}, \mathbf{v}](\mathbf{n}, \mathbf{x})}{(\mathbf{n}, \operatorname{curl}\mathbf{n})}. \tag{13}$$

In the case when $\mathbf{x}$ is orthogonal to $\mathbf{n}$ we shall come to the equation of a parallel-transportable field along the admissible curve introduced above.

We shall call the parallel transport defined by (13) the *prolongated parallel transport*. It is easy to check that if the scalar product $(\mathbf{n}, \mathbf{v})$ is zero at some initial point on the curve of parallel transport then by virtue of (13) this product is zero identically. This means that the parallel transport takes a tangent to the non-holonomic manifold vector into the tangent, as well.

Equation (13) is a consequence of equation (3). Conversely, equation (13) implies (3). Therefore, (13) is equivalent to (3). But (13) is preferable to (3) because it contains the operator $d\mathbf{v}/dr$ which acts over an arbitrary vector field $\mathbf{x}$.

The equivalence of (3) and (13) means that if there is a parallel-transportable vector field in the sense of (3) then it will be parallel-transportable in sense of (13), i.e. with respect to prolongated parallel transport. Find the integrability condition of equation (13). Let $\mathbf{x}$ and $\mathbf{y}$ be constant vector fields in E^3. Differentiating both sides of (3) in the direction of the vector $\mathbf{y}$ we obtain

$$\begin{aligned}\frac{d^2\mathbf{v}}{dr^2}(\mathbf{x}, \mathbf{y}) = &- A\mathbf{y}(\mathbf{v}, A\mathbf{x}) - \mathbf{n}\left(\frac{d\mathbf{v}}{dr}\mathbf{y}, A\mathbf{x}\right) - \mathbf{n}\left(\mathbf{v}, \frac{d^2\mathbf{n}}{dr^2}(\mathbf{x}, \mathbf{y})\right) \\ &- [\mathbf{n}, \mathbf{v}]\left(\mathbf{x}, \frac{d}{dr}\left(\frac{K\mathbf{n}}{(\mathbf{n}, \operatorname{curl}\mathbf{n})}\right)\mathbf{y}\right) - \frac{K(\mathbf{n}, \mathbf{x})}{(\mathbf{n}, \operatorname{curl}\mathbf{n})}\left([A\mathbf{y}, \mathbf{v}] + \left[\mathbf{n}, \frac{d\mathbf{v}}{dr}\mathbf{y}\right]\right).\end{aligned} \tag{14}$$

Make a substitution of $d\mathbf{v}/dr$ by means of (13). Also, using the orthogonality of $\mathbf{v}$ and $A\mathbf{y}$ to the field $\mathbf{n}$, we can write

$$[A\mathbf{y}, \mathbf{v}] = \mathbf{n}(\mathbf{n}, A\mathbf{y}, \mathbf{v}).$$

Equation (14) can be rewritten as

$$\frac{d^2\mathbf{v}}{dr^2}(\mathbf{x}, \mathbf{y}) = -A\mathbf{y}(\mathbf{v}, A\mathbf{x}) - [\mathbf{n}, \mathbf{v}]\left(\mathbf{x}, \frac{d}{dr}\left(\frac{K\mathbf{n}}{(\mathbf{n}, \operatorname{curl}\mathbf{n})}\right)\mathbf{y}\right) - \mathbf{n}\left(\mathbf{v}, \frac{d^2\mathbf{n}}{dr^2}(\mathbf{x}, \mathbf{y})\right)$$
$$+ \frac{K\mathbf{n}}{(\mathbf{n}, \operatorname{curl}\mathbf{n})}\left\{(\mathbf{n}, \mathbf{x})(\mathbf{n}, \mathbf{v}, A\mathbf{y}) + (\mathbf{n}, \mathbf{y})(\mathbf{n}, \mathbf{v}, A\mathbf{x})\right\}$$
$$+ \left(\frac{K\mathbf{n}}{(\mathbf{n}, \operatorname{curl}\mathbf{n})}\right)^2 (\mathbf{n}, \mathbf{x})(\mathbf{n}, \mathbf{y})[\mathbf{n}, [\mathbf{n}, \mathbf{v}]].$$

The last three terms on the right of the equation are symmetric with respect to $\mathbf{x}$ and $\mathbf{y}$. Interchanging the roles of $\mathbf{x}$ and $\mathbf{y}$ and subtracting the equation obtained from the original, we get

$$[\mathbf{n}, \mathbf{v}]\left(\mathbf{x}, \mathbf{y}, \operatorname{curl}\frac{K\mathbf{n}}{(\mathbf{n}, \operatorname{curl}\mathbf{n})}\right) + [\mathbf{v}, [A\mathbf{x}, A\mathbf{y}]] = 0. \tag{15}$$

Consider the second term. Transform the vector product in it as

$$[A\mathbf{x}, A\mathbf{y}] = \mathbf{n}(\mathbf{n}, A\mathbf{x}, A\mathbf{y}). \tag{16}$$

With respect to the special system of coordinates defined above, we obtain

$$(\mathbf{n}, A\mathbf{x}, A\mathbf{y}) = \begin{vmatrix} \xi_i^1 x^i & \xi_i^2 x^i & 0 \\ \xi_i^1 y^i & \xi_i^2 y^i & 0 \\ 0 & 0 & 1 \end{vmatrix}$$
$$= \begin{vmatrix} \xi_2^1 & \xi_2^2 \\ \xi_3^1 & \xi_3^2 \end{vmatrix}\begin{vmatrix} x^2 & x^3 \\ y^2 & y^3 \end{vmatrix} + \begin{vmatrix} \xi_3^1 & \xi_3^2 \\ \xi_1^1 & \xi_1^2 \end{vmatrix}\begin{vmatrix} x^3 & x^1 \\ y^3 & y^1 \end{vmatrix} + \begin{vmatrix} \xi_1^1 & \xi_1^2 \\ \xi_2^1 & \xi_2^2 \end{vmatrix}\begin{vmatrix} x^1 & x^2 \\ y^1 & y^2 \end{vmatrix}.$$

The latter expression is the scalar product of the curvature vector $\mathbf{P}$ (see Section 1.5) and $[\mathbf{x}, \mathbf{y}]$. In Section 1.9 we stated the invariant form of the curvature vector

$$\mathbf{P} = K\mathbf{n} + H\mathbf{k} + \nabla_{\mathbf{k}}\mathbf{n}.$$

So,

$$(\mathbf{n}, A\mathbf{x}, A\mathbf{y}) = (\mathbf{P}, \mathbf{x}, \mathbf{y}). \tag{17}$$

Turn back to (15). Substitute it into (16) and (17). We get

$$[\mathbf{n}, \mathbf{v}]\left(\mathbf{x}, \mathbf{y}, \operatorname{curl}\left(\frac{K\mathbf{n}}{(\mathbf{n}, \operatorname{curl}\mathbf{n})}\right) - \mathbf{P}\right) = 0.$$

But $[\mathbf{n}, \mathbf{v}] \neq 0$ and $\mathbf{x}, \mathbf{y}$ are arbitrary vectors. Therefore,

$$\operatorname{curl}\left(\frac{K\mathbf{n}}{(\mathbf{n}, \operatorname{curl}\mathbf{n})}\right) - \mathbf{P} = 0.$$

We shall call the vector on the left the *Vagner vector*. So, the following theorem holds.

Theorem *In order that the vector field parallel-transportable on the whole non-holonomic manifold exists it is necessary and sufficient that the Vagner vector*

$$\operatorname{curl}\left(\frac{K\mathbf{n}}{(\mathbf{n}, \operatorname{curl}\mathbf{n})}\right) - (K\mathbf{n} - \mathbf{k}\operatorname{div}\mathbf{n} + \nabla_{\mathbf{k}}\mathbf{n})$$

is zero.

Give an example of a vector field having a zero Vagner vector. Define the components of that field as

$$\xi^1 = \cos\varphi, \quad \xi^2 = \sin\varphi, \quad \xi^3 = 0,$$

where $\varphi = cx^3 + d$, $c \neq 0$ and d are the constants. In Section 1.1 we found that the non-holonomicity value of that field is $(\mathbf{n}, \operatorname{curl}\mathbf{n}) = -c$. As the field $\mathbf{n}$ streamlines are the straight lines, the vector of the curvature of the streamlines $\mathbf{k} = 0$. Hence $\nabla_{\mathbf{k}} = 0$. The field $\mathbf{n}$ is parallel to the x^1x^2-plane. Therefore, the mapping ψ maps any surface into the arc of the same great circle in the unit sphere. Hence, $K = 0$. We see that all the terms in the expression for the Vagner vector become zero.

2 Vector Fields and Differential Forms in Many-Dimensional Euclidean and Riemannian Spaces

2.1 The Unit Vector Field in Many-Dimensional Euclidean Space

Let us given a unit vector field **n** in some domain G of $(m+1)$-dimensional Euclidean space E^{m+1}. For any shift $d\mathbf{r}$ from $M \in G$ in the direction orthogonal to $\mathbf{n}(M)$ we define the normal curvature of the first kind k_n setting $k_n = -(d\mathbf{n}, d\mathbf{r})/d\mathbf{r}^2$.

We shall call the extremal values of the first kind of normal curvature the *principal curvature of the first kind*. We shall call the directions $d\mathbf{r}$ for which these values are achieved *the principal directions of the first kind*. Let us find the system of equations to determine them from. Introduce the Cartesian system of coordinates x^i in E^{m+1}. We denote the vector components with respect to this system with superscripts; for instance, the components of **n** will be denoted by ξ^i. To simplify notations, we denote the derivative $\partial\xi^i/\partial x^j$ by ξ^i_j. Consider the field **n** in the neighborhood of M. Choose the Cartesian coordinates in such a way that the basis vector $\mathbf{e}_{m+1}$ will be directed along $\mathbf{n}(M)$. The fact that $d\mathbf{r}$ is orthogonal to $\mathbf{n}(M)$ implies $d\mathbf{r} = \{dx^1, \dots, dx^m, 0\}$. In solving the extremal problem we come to the system of principal directions:

$$
\begin{array}{ccccccccc}
(\xi^1_1 + \lambda)\, dx^1 & + & \dfrac{\xi^1_2 + \xi^2_1}{2} dx^1 & + & \cdots & + & \dfrac{\xi^1_m + \xi^m_1}{2} dx^m & = 0, \\
\dfrac{\xi^1_2 + \xi^2_1}{2} dx^1 & + & (\xi^2_2 + \lambda)\, dx^2 & + & \cdots & + & & = 0, \\
\dots & \dots & \dots & \dots & \dots & \dots & & \\
\dfrac{\xi^1_m + \xi^m_1}{2} dx^1 & + & \cdots & + & \cdots & + & (\xi^m_m + \lambda)\, dx^m & = 0.
\end{array}
$$

The principal curvatures of the first kind are the roots of the characteristic polynomial

$$\left|\lambda\delta_{ik} + \frac{\xi^i_k + \xi^k_i}{2}\right| = 0. \tag{1}$$

As the matrix $\|(\xi^i_k + \xi^k_i)/2\|$ is symmetric, all the roots λ_i are real. Denote by σ_k the k-th symmetric polynomial of the principal curvatures of the first kind. The σ_k can be expressed in terms of the main minors of the matrix $\|(\xi^i_k + \xi^k_i)/2\|$ as follows:

$$\sigma_k = (-1)^k \sum_{1\le i_1<i_2<\cdots<i_k}^{m} \begin{vmatrix} \xi^{i_1}_{i_1} & \dots & \frac{\xi^{i_1}_{i_k}+\xi^{i_k}_{i_1}}{2} \\ \dots & \dots & \dots \\ \frac{\xi^{i_1}_{i_k}+\xi^{i_k}_{i_1}}{2} & \dots & \xi^{i_k}_{i_k} \end{vmatrix}. \tag{2}$$

We define the principal curvatures and principal directions of the second kind by means of the Rodrigues equation

$$d\mathbf{n} = -\mu\, d\mathbf{r}. \tag{3}$$

We call the direction $d\mathbf{r}$ satisfying (3) and orthogonal to $\mathbf{n}$ the *principal direction of the second kind*.

We can find the number μ, which satisfies the Rodrigues equation, from the following equation:

$$\begin{vmatrix} \xi^1_1+\mu & \xi^1_2 & \dots & \xi^1_{m+1} \\ \xi^2_1 & \xi^2_2+\mu & \dots & \dots \\ \dots & \dots & \dots & \dots \\ \xi^{m+1}_1 & \dots & \dots & \xi^{m+1}_{m+1}+\mu \end{vmatrix} = 0. \tag{4}$$

Here we do not suppose a special choice of space coordinate system. One of the roots $\mu_1, \dots, \mu_{m+1}$ of equation (4) is zero. Indeed, by means of the vector field $\mathbf{n}$ we are able to construct the mapping ψ of the $(m+1)$-dimensional domain G onto the m-dimensional domain G in the unit sphere S^m. Therefore, the Jacobian of ψ, which is equal to the determinant of $\|\xi^i_j\|$, is equal to zero. Hence, one of the roots of (4) is zero. For definiteness, set $\mu_{m+1} = 0$. The other roots are called, by definition, the *principal curvatures of the second kind*. Since the matrix $\|\xi^i_j\|$ is not symmetric in general, the μ_i can be complex. Therefore, the real principal directions do not exist for these μ_i. But in all cases the symmetric polynomials of the μ_i are real ones. They can be expressed in terms of coefficients of the polynomial (4).

Denote the symmetric polynomial of the principal curvatures of the second kind by S_k. It is easy to express them in terms of derivatives of components of the vector field $\mathbf{n}$:

$$S_k = (-1)^k \sum_{1\le i_1<i_2<\cdots<i_k}^{m+1} \begin{vmatrix} \xi^{i_1}_{i_1} & \dots & \xi^{i_1}_{i_k} \\ \dots & \dots & \dots \\ \xi^{i_k}_{i_1} & \dots & \xi^{i_k}_{i_k} \end{vmatrix}, \quad k = 1, \dots, m+1.$$

Observe that $\sigma_1 = S_1$ and $S_{m+1} = 0$ because of $\mu_{m+1} = 0$.

2.2 The Regular Vector Field Defined in a Whole Space

We shall consider the unit continuous vector fields defined at all points of E^{m+1} including a point at infinity. We can put every such field into correspondence with a regular mapping of the unit sphere S^{m+1} onto the unit sphere S^m and vice versa. Indeed, if we regard E^{m+1} as a tangent space of S^{m+1} at the north pole then by means of stereographic projection from the south pole we are able to map S^{m+1} onto E^{m+1}. The south pole corresponds to a point at infinity in E^{m+1}. Let $p \in S^{m+1}$ and x be the corresponding point in E^{m+1} in stereographic projection. Now construct a mapping ψ of Euclidean space E^{m+1} onto S^m. To do this, we put $\mathbf{n}(x)$ into the origin $O \in E^{m+1}$. The end-point of $\mathbf{n}(x)$ then marks a point q in S^m. Put into correspondence the points $p \in S^{m+1}$ and $q \in S^m$. Since $\mathbf{n}(x)$ is defined at all points of E^{m+1} including a point at infinity, the mapping of S^{m+1} onto S^m is defined at all points of S^{m+1} including the south pole. Conversely, by a given mapping of S^{m+1} onto S^m we are able to construct a vector field in E^{m+1}.

The Freudenthal theorem on the classification of mappings $S^{m+1} \to S^m$ is known: the homotopy group $\pi_{m+1}(S^m) = \mathbf{Z}_2$. In Pontryagin's book (see [72], p. 154) the following theorem has been proved. *For $m \geq 3$ there are exactly two homotopic classes of mappings $S^{m+1} \to S^m$.* Whether or not the mapping $f: S^{m+1} \to S^m$ belongs to one of those classes depends on some invariant $\delta(f)$ which is the residue modulo 2 and therefore takes only values 0, 1. The construction of this invariant is as follows. Let $q \in S^m$ be a regular point of the mapping f. Then $f^{-1}(q)$ is a one-dimensional submanifold in S^{m+1}. Denote by M^1 the image of this submanifold in a stereographic projection of S^{m+1} onto E^{m+1}. In terms of a vector field the submanifold M^1 consists of one or several closed curves where the field $\mathbf{n}$ is constant. If the set $f^{-1}(q)$ does not contain the south pole then M^1 is a smooth closed submanifold of Euclidean space E^m. By means of the mapping $\psi: E^{m+1} \to S^m$ we are able to frame M^1 with the fields orthogonal to M^1. To do this we consider the tangent space T_q at $q \in S^m$. The mapping ψ induces the linear mapping ψ_* of E^{m+1} onto T_q at $x \in M^1$. Choose the basis $\mathbf{e}_1, \dots, \mathbf{e}_m$ in T_q. We take the vectors $\mathbf{u}_i = \psi_*^{-1}\mathbf{e}_i$ as a framing of M_1 at x. By means of a continuous deformation, this framing can be transformed into both an orthonormal and normal to an M^1 framing. We denote the vector fields of the latter framing along M^1 by $\mathbf{u}_1, \dots, \mathbf{u}_m$. We denote the vector field tangent to M^1 by $\mathbf{u}_{m+1}$. Thus, for the mapping $f: S^{m+1} \to S^m$ we can put into correspondence some framed one-dimensional closed submanifold M^1.

Let $\tau_1, \dots, \tau_{m+1}$ be some positively oriented orthonormal basis in E^{m+1}. Then

$$\mathbf{u}_i(x) = \sum_{j=1}^{m+1} h_{ij}(x)\tau_j, \quad i = 1, \dots, m+1,$$

where $\|h_{ij}(x)\|$ is an orthogonal matrix with a positive determinant. The field of those matrices along M^1 defines a continuous mapping h of the curve M^1 into the manifold H_{m+1} of all rotations of Euclidean space E^{m+1}. Now we define the invariant $\beta(h)$ — the residue modulo 2. If $m \geq 2$ then for the unicomponent manifold M^1 the residue $\beta(h) = 0$ if h is homotopy equivalent in H_{m+1} to a point and $\beta(h) = 1$ otherwise. For the case of a many-component manifold the residue $\beta(h)$ is equal to the sum of each of the component residues. If $m = 1$ then $\beta(h)$ is a degree modulo 2 mapping of M^1 onto the circle H_2. Next, let $r(M^1)$ be the number of components of M^1. Then the invariant $\delta(f)$ of the mapping f and, consequently, the field **n** invariant under continuous deformations is the number

$$\delta(f) \equiv \beta(h) + r(M^1) \pmod 2.$$

Consider the unit field **n** defined in Euclidean space E^{n+r} including a point at infinity which is parallel to some space E_0^{n+1}. By means of a stereographic projection we are able to state the one-to-one correspondence between the points of S^{n+r} and E^{n+r}. Therefore, the field **n** generates a mapping $f\colon S^{n+r} \to S^n$. The homotopies of f correspond to those deformations of **n** in E^{n+r} which preserve the property of **n** to be parallel to E_0^{n+1}.

The classes of continuous mappings f form the homotopy group $\pi_{n+r}(S^n)$. Some element of this homotopy group corresponds to the field **n**. The problem of finding all the homotopy groups of spheres is an unsolved one. The solution was given for r not very large.

Pontryagin stated that $\pi_{n+2}(S^n) = \mathbf{Z}_2$ *and pointed out the method of evaluating the invariant which classifies the mappings* $S^{n+2} \to S^n$ *up to homotopies*. We apply that method to the vector field defined in Euclidean space E^{n+2}. Let $Q \in S^n$ be a regular point. Then M^2 — the preimage of Q — is the two-dimensional surface in E^{n+2} on which the field **n** is constant and the end-point of **n** coincides with Q when the initial point of **n** is put to the center of S^n. If M^2 consists of a single component and is a surface of genus p then it is possible to select a set of smooth closed curves $A_1, \dots, A_p, B_1, \dots, B_p$ such that A_i intersects B_i at a single point avoiding tangency while any other curves have no intersections. For any curve C of this set we are able to determine the invariant $\delta(C)$. To do this we shall construct a framing of C. Set $\mathbf{u}_{n+1}$ to be the vector which is normal to C and tangent to M^2. The surface M^2 is framed with normal fields $\mathbf{u}_1, \dots, \mathbf{u}_n$ which are induced by mapping onto S^n. Therefore, C is framed with $\mathbf{u}_1, \dots, \mathbf{u}_n, \mathbf{u}_{n+1}$. Let τ be the tangent vector of C. Decomposing the vectors $\mathbf{u}_1, \dots, \mathbf{u}_{n+1}, \tau$ with respect to the fixed orthonormal basis in E^{n+2}, we obtain as above the mapping of C into the group H_{n+2} of orthogonal matrices of order $n+2$. If this mapping is homotopic to the constant one then $\delta(C) = 1$, otherwise $\delta(C) = 0$. Then the Pontryagin invariant is a residue modulo 2:

$$\delta(M^2) = \sum_{i=1}^{p} \delta(A_i)\delta(B_i).$$

If M^2 consists of several components then $\delta(M^2)$ is the sum of its values on each component.

As an example, evaluate the δ invariant for the vector field **n** defined in E^4 and parallel to E_0^3. The field **n** has the following components:

$$\xi_1 = \frac{2a}{1+a^2+b^2}, \quad \xi_2 = \frac{2b}{1+a^2+b^2}, \quad \xi_3 = \frac{a^2+b^2-1}{1+a^2+b^2}, \quad \xi_4 = 0,$$

where $a = \sqrt{x_1^2 + x_2^2}$, $b = \sqrt{x_3^2 + x_4^2}$. Set $\mathbf{e}_1, \mathbf{e}_2, \mathbf{e}_3, \mathbf{e}_4$ to be a fixed basis in E^4. Evidently, we can represent the Cartesian coordinates in E^4 in the following form:

$$x_1 = a\cos\alpha, \quad x_2 = a\sin\alpha, \quad x_3 = b\cos\beta, \quad x_4 = b\sin\beta.$$

We get the surface M^2 on which the field $\mathbf{n}$ is constant taking a and b fixed. Hence, all of the surfaces M^2 for $a \neq 0, b \neq 0$ are the Clifford tori. Let A_1 be a curve defined by $\beta = \text{const}$, while B_2 is a curve defined by $\alpha = \text{const}$. Construct the framing of the curves generated by the field $\mathbf{n}$. The vectors $\mathbf{n}_a$ and $\mathbf{n}_b$ (the derivatives of $\mathbf{n}$ in parameters a and b respectively) are tangent to the sphere S^2. Consider the expansion of the field $\mathbf{n}$ nearby $x_0 \in M^2$:

$$\mathbf{n}(x) = \mathbf{n}(x_0 + \Delta x) = \mathbf{n}(x_0) + \left(\mathbf{n}_a\frac{\partial a}{\partial x_i} + \mathbf{n}_b\frac{\partial b}{\partial x_i}\right)\Delta x_i$$

$$= \mathbf{n}(x_0) + \mathbf{n}_a(\cos\alpha\,\Delta x_1 + \sin\alpha\,\Delta x_2) + \mathbf{n}_b(\cos\beta\,\Delta x_3 + \sin\beta\,\Delta x_4).$$

Here we have used the following expressions:

$$\frac{\partial a}{\partial x_1} = \cos\alpha, \quad \frac{\partial a}{\partial x_2} = \sin\alpha, \quad \frac{\partial a}{\partial x_3} = 0, \quad \frac{\partial a}{\partial x_4} = 0,$$

$$\frac{\partial b}{\partial x_1} = 0, \quad \frac{\partial b}{\partial x_2} = 0, \quad \frac{\partial b}{\partial x_3} = \cos\beta, \quad \frac{\partial b}{\partial x_4} = \sin\beta.$$

Under the induced mapping of the tangent space of E^4 at x_0 the framing vectors $\mathbf{u}_1(x)$ and $\mathbf{u}_2(x)$ must be turned into the vectors $\mathbf{n}_a$ and $\mathbf{n}_b$ respectively. Consider the torus normals

$$\mathbf{n}_1 = \cos\alpha\mathbf{e}_1 + \sin\alpha\mathbf{e}_2, \quad \mathbf{n}_2 = \cos\beta\mathbf{e}_3 + \sin\beta\mathbf{e}_4.$$

Let the shift Δx be directed along $\mathbf{n}_1$. Then $d\mathbf{n}$ is proportional to $\mathbf{n}_a$, namely $d\mathbf{n} = \mathbf{n}_a\,da$. If Δx is directed along $\mathbf{n}_2$ then $d\mathbf{n} = \mathbf{n}_b\,db$. Hence, the framing vectors $\mathbf{u}_1$ and $\mathbf{u}_2$ coincide with $\mathbf{n}_1$ and $\mathbf{n}_2$ respectively. Next, the vector $\tau_1 = -\sin\alpha\mathbf{e}_1 + \cos\alpha\mathbf{e}_2$ is tangent to A_1, while $\tau_2 = -\sin\beta\mathbf{e}_3 + \cos\beta\mathbf{e}_4$ is tangent to M^2 and orthogonal to A_1. The matrix of transformation of the fixed basis $\mathbf{e}_1, \mathbf{e}_2, \mathbf{e}_3, \mathbf{e}_4$ into the basis $\mathbf{u}_1, \mathbf{u}_2, \tau_1, \tau_2$ has the form

$$h(x) = \begin{Vmatrix} \cos\alpha & \sin\alpha & 0 & 0 \\ 0 & 0 & \cos\beta & \sin\beta \\ -\sin\alpha & \cos\alpha & 0 & 0 \\ 0 & 0 & -\sin\beta & \cos\beta \end{Vmatrix}.$$

On the curve A_1 we have $\beta = \text{const}$, while α varies from 0 to 2π. The theorem has been stated (see [72], p. 132) which allows a check to be made as to whether or not a given mapping h of the circle S^1 into the group H_n of orthogonal matrices is homotopic to the constant one. Every mapping $h\colon S^1 \to H_n$ is continuously homotopic to a mapping $g\colon S^1 \to H_2 = S^1$. It happens that g is homotopic to the constant mapping in H_n $(n \geq 3)$ if and only if the degree of g is even.

In our case the mapping h by itself has the form of $g: S^1 \to H_2$. This mapping defines a rotation of $\mathbf{u}_1, \tau_1$ in the $\mathbf{e}_1, \mathbf{e}_2$-plane. The degree of g is 1. Therefore, $g: A_1 \to H_4$ is not homotopic to the constant mapping and by definition $\delta(A_1) = 0$. So, $\delta(M^2) = 0$. It is easy to check directly that the field $\mathbf{n}(M)$ is homotopic to the constant one. To do this, replace a with ta, b with tb in the expressions for ξ_i. Then $t = 1$ produces the given field $\mathbf{n}$, while $t = 0$ produces the constant field.

The groups $\pi_{n+r}(S^n)$ for $r \geq 15$ have been found by V. Rohlin, J.-P. Serre, A. Cartan, Toda and others. The variety of the structure of these groups and, on the other hand, their stabilization in growth of $\mathbf{n}$ became clear. Here we cite some results on homotopy groups [78]:

$$\pi_{n+3} = \mathbf{Z}_{24} \text{ for } n \geq 5; \quad \pi_{n+4} = 0 \text{ for } n \geq 6;$$
$$\pi_{n+5} = 0 \text{ for } n \geq 7; \quad \pi_{n+6} = \mathbf{Z}_2 \text{ for } n \geq 5;$$
$$\pi_{n+7} = \mathbf{Z}_{240} \text{ for } n \geq 9;$$

The latter isomorphism is a most impressive result.

2.3 The Many-Dimensional Generalization of the Gauss–Bonnet Formula to the Case of a Vector Field

Consider the unit vector field defined in some domain G of E^{m+1}. We suppose that $\mathbf{n}$ is regular of class C^1 except for some points, maybe. Let F^m be a closed hypersurface in E^{m+1} which does not pass through the singular points of the field $\mathbf{n}$. Denote the curvilinear coordinates in some domain of the hypersurface F^m by $\alpha_1, \ldots, \alpha_m$. By means of the vector field $\mathbf{n}$ we can construct the mapping ψ of this domain onto the unit sphere S^m with the center at the origin 0. To do this we put into correspondence the point $M \in F^m$ and the point in S^m which is the end-point of $\mathbf{n}(M)$ being taken to the origin. The field $\mathbf{n}$ depends on $\alpha_1, \ldots, \alpha_m$ along F^m and hence they can be considered as the coordinates in S^m, as well. Denote by $d\sigma$ the m-volume element of S^m. Then

$$d\sigma = (\mathbf{n}_{\alpha_1}, \ldots, \mathbf{n}_{\alpha_m}, \mathbf{n})\, d\alpha_1 \ldots d\alpha_m, \tag{1}$$

where the parentheses mean the mixed product of the $(m+1)$ vector set in E^{m+1}. Since the field $\mathbf{n}$ components are regular functions of Cartesian coordinates x^i, then by the composite function differentiation rule we have

$$(\mathbf{n}_{\alpha_1}, \ldots, \mathbf{n}_{\alpha_m}, \mathbf{n}) = \sum_{i_1, \ldots, i_m = 1}^{m+1} (\mathbf{n}_{x^{i_1}}, \ldots, \mathbf{n}_{x^{i_m}}, \mathbf{n}) \frac{\partial x^{i_1}}{\partial \alpha_1} \cdots \frac{\partial x^{i_m}}{\partial \alpha_m}. \tag{2}$$

For the sake of convenience, denote the derivative $\mathbf{n}_{x^i}$ by $\mathbf{n}_i$. If the set of indices $j_1, \ldots, j_m$ differs from the set $i_1, \ldots, i_m$ only in their order then

$$(\mathbf{n}_{i_1}, \ldots, \mathbf{n}_{i_m}, \mathbf{n}) = \varepsilon^{j_1 \ldots j_m}_{i_1 \ldots i_m} (\mathbf{n}_{j_1}, \ldots, \mathbf{n}_{j_m}, \mathbf{n}) \tag{3}$$

where $\varepsilon^{j_1 \dots j_m}_{i_1 \dots i_m}$ is the Kronecker symbol. Consider the following set of index strings

$$(2, 3, \dots, m+1),$$
$$(1, 3, \dots, m+1),$$
$$\dots$$
$$(1, 2, \dots, m).$$

Denote this set by Λ. Each of these strings consists of m distinct integers from the set $1, 2, \dots, m+1$. Let $(j_1, \dots, j_m) \in \Lambda$. On the right of (2) we rearrange the terms of the sum in such a way that each string of indices $(i_1, \dots, i_m)$ would be different from the fixed string $(j_1, \dots, j_m)$ only in their order. Then this sum obtains the form

$$(\mathbf{n}_{j_1}, \dots, \mathbf{n}_{j_m}, \mathbf{n}) \sum_{i_1, \dots, i_m}^{m+1} \varepsilon^{j_1 \dots j_m}_{i_1 \dots i_m} \frac{\partial x^{i_1}}{\partial \alpha_1} \cdots \frac{\partial x^{i_m}}{\partial \alpha_m}.$$

The multiplier of $(\mathbf{n}_{j_1}, \dots, \mathbf{n}_{j_m}, \mathbf{n})$ is the determinant

$$I_{j_1 \dots j_m} = \begin{vmatrix} \frac{\partial x^{j_1}}{\partial \alpha_1} & \dots & \frac{\partial x^{j_m}}{\partial \alpha_1} \\ \dots & \dots & \dots \\ \frac{\partial x^{j_1}}{\partial \alpha_m} & \dots & \frac{\partial x^{j_m}}{\partial \alpha_m} \end{vmatrix}.$$

Using that notation, we can write

$$\begin{aligned}(\mathbf{n}_{\alpha_1}, \dots, \mathbf{n}_{\alpha_m}, \mathbf{n}) = {} & (\mathbf{n}_2, \mathbf{n}_3, \dots, \mathbf{n}_{m+1}, \mathbf{n}) I_{23\dots m+1} \\ & + (\mathbf{n}_1, \mathbf{n}_3, \dots, \mathbf{n}_{m+1}, \mathbf{n}) I_{13\dots m+1} + \cdots \\ & \cdots + (\mathbf{n}_1, \mathbf{n}_2, \dots, \mathbf{n}_m, \mathbf{n}) I_{12\dots m}. \end{aligned} \tag{4}$$

The values $I_{j_1, \dots, j_m}$ can be expressed in terms of the F^m normal vector components. If $\mathbf{r}(\alpha_1, \dots, \alpha_m)$ is a position vector of F^m then the unit normal ν is the normalized vector product of $\mathbf{r}_{\alpha_1}, \dots, \mathbf{r}_{\alpha_m}$:

$$\nu = \frac{[\mathbf{r}_{\alpha_1}, \dots, \mathbf{r}_{\alpha_m}]}{|[\mathbf{r}_{\alpha_1}, \dots, \mathbf{r}_{\alpha_m}]|}. \tag{5}$$

Let $\mathbf{e}_1, \dots, \mathbf{e}_{m+1}$ be a positively oriented orthonormal basis in E^{m+1}. We assume the vector product such that $[\mathbf{e}_1, \dots, \mathbf{e}_m] = \mathbf{e}_{m+1}$. Then the vector product can be found in terms of the signed determinant, namely

$$[\mathbf{r}_{\alpha_1}, \dots, \mathbf{r}_{\alpha_m}] = (-1)^m \begin{vmatrix} \mathbf{e}_1 & \dots & \mathbf{e}_{m+1} \\ \frac{\partial x^1}{\partial \alpha_1} & \dots & \frac{\partial x^{m+1}}{\partial \alpha_1} \\ \dots & \dots & \dots \\ \frac{\partial x^1}{\partial \alpha_m} & \dots & \frac{\partial x^{m+1}}{\partial \alpha_m} \end{vmatrix}.$$

From this we find the components of the F^m unit normal. By means of (4) and (5), we obtain

$$\begin{aligned}(\mathbf{n}_{\alpha_1}, \dots, \mathbf{n}_{\alpha_m}, \mathbf{n}) = {} & (-1)^m \{ (\mathbf{n}_2, \mathbf{n}_3, \dots, \mathbf{n}_{m+1}, \mathbf{n}) \nu_1 \\ & - (\mathbf{n}_1, \mathbf{n}_3, \dots, \mathbf{n}_{m+1}, \mathbf{n}) \nu_2 + \cdots \\ & \cdots + (-1)^m (\mathbf{n}_1, \mathbf{n}_2, \dots, \mathbf{n}_m, \mathbf{n}) \nu_{m+1} \} |[\mathbf{r}_{\alpha_1}, \dots, \mathbf{r}_{\alpha_m}]|. \end{aligned} \tag{6}$$

Note that the m-volume element dV of F^m has the following expression

$$dV = \|[\mathbf{r}_{\alpha_1}, \dots, \mathbf{r}_{\alpha_m}]\|\, d\alpha_1 \dots d\alpha_m. \tag{7}$$

Introduce the vector field $\mathbf{P}$ as follows:

$$\begin{aligned}\mathbf{P} = (-1)^m\{&(\mathbf{n}_2, \mathbf{n}_3, \dots, \mathbf{n}_{m+1}, \mathbf{n}), -(\mathbf{n}_1, \mathbf{n}_3, \dots, \mathbf{n}_{m+1}, \mathbf{n}), \dots,\\ &(-1)^m(\mathbf{n}_1, \mathbf{n}_2, \dots, \mathbf{n}_m, \mathbf{n})\}.\end{aligned} \tag{8}$$

From (1), (6)–(8) the formula for the volume element of the unit sphere S^m follows, namely

$$d\sigma = (\mathbf{P}, \nu)\, dV. \tag{9}$$

Here the parentheses mean the scalar product in E^{m+1}. If F^m is a closed hypersurface then the ψ image of F^m covers the unit sphere some integer number of times. More precisely, we are able to put the mapping ψ into correspondence with the integer, namely, the degree of mapping ψ (see for definitions [78]). This number can be found by the formula

$$\theta = \frac{1}{\omega_m} \int\limits_{\psi(F^m)} d\sigma,$$

where the integral expresses the volume of the image of the mapping ψ if both the sign of $d\sigma$ and the covering multiplicity of the domains in S^m are taken into account. The ω_m stands for the volume of the unit sphere S^m. Integrating (8) over the hypersurface F^m, we obtain

$$\int\limits_{F^m} (\mathbf{P}, \nu)\, dV = \omega_m \theta.$$

We call the vector $\mathbf{P}$ the curvature vector of the field $\mathbf{n}$. This vector $\mathbf{P}$ is defined with respect to any Cartesian system of coordinates. Now we state the invariant representation of $\mathbf{P}$ which is analogous to one in Section 1.10. *Define the vector sequence* $\mathbf{k}_i$ *inductively*:

$$\mathbf{k}_0 = \mathbf{n}, \dots, \mathbf{k}_{i+1} = \nabla_{\mathbf{k}_i}\mathbf{n}, \dots,$$

where $i = 0, 1, \dots m$. Remember that we denoted by S_i the symmetric polynomials of the principal curvatures of the second kind. The following theorem holds.

Theorem *The curvature vector* $\mathbf{P}$ *can be expressed in terms of* $\mathbf{k}_i$ *and symmetric polynomials* S_i *with the following formula:*

$$\mathbf{P} = (-1)^m\{S_m\mathbf{n} + S_{m-1}\mathbf{k}_1 + \dots + \mathbf{k}_m\}. \tag{10}$$

There is another way to represent $\mathbf{k}_i$. Let $A = \|a_{ij}\|$ be the matrix of order $(m+1)$:

$$A = \begin{Vmatrix} \xi_1^1 & \dots & \xi_{m+1}^1 \\ \dots & \dots & \dots \\ \xi_1^{m+1} & \dots & \xi_{m+1}^{m+1} \end{Vmatrix}.$$

The vector $\mathbf{k}_1$ has components $\xi_p^\alpha \xi^p$, where the summation over p is assumed. Therefore, the components of $\mathbf{k}_2$ are as follows: $\xi_l^\alpha \xi_p^l \xi^p$. In general, the components of $\mathbf{k}_i$ are

$$\xi_j^\alpha \xi_r^j \xi_s^r \ldots \xi_p^l \xi^p,$$

where the summation over $j, r, s, \ldots, l, p$ is assumed and the number of indices under summation is i. The index α shows that the expression stands for the α-th component of $\mathbf{k}_i$. On the other hand, the vector $A\mathbf{n}$ has the form

$$A\mathbf{n} = \begin{Vmatrix} \xi_1^1 & \cdots & \xi_{m+1}^1 \\ \cdots & \cdots & \cdots \\ \xi_1^{m+1} & \cdots & \xi_{m+1}^{m+1} \end{Vmatrix} \begin{pmatrix} \xi^1 \\ \vdots \\ \xi^{m+1} \end{pmatrix} = \begin{pmatrix} \xi_p^1 \xi^p \\ \vdots \\ \xi_p^{m+1} \xi^p \end{pmatrix}.$$

Comparing the components of $\mathbf{k}_1$ and $A\mathbf{n}$, we see that $\mathbf{k}_1 = A\mathbf{n}$. Further, $A^2\mathbf{n} = AA\mathbf{n}$ has the form

$$A^2\mathbf{n} = \begin{Vmatrix} \xi_1^1 & \cdots & \xi_{m+1}^1 \\ \cdots & \cdots & \cdots \\ \xi_1^{m+1} & \cdots & \xi_{m+1}^{m+1} \end{Vmatrix} \begin{pmatrix} \xi_p^1 \xi^p \\ \vdots \\ \xi_p^{m+1} \xi^p \end{pmatrix} = \begin{pmatrix} \xi_l^1 \xi_p^l \xi^p \\ \vdots \\ \xi_l^{m+1} \xi_p^l \xi^p \end{pmatrix}.$$

Comparing components, we again conclude that $\mathbf{k}_2$ and $A^2\mathbf{n}$ coincide. In an analogous manner we find that $\mathbf{k}_i = A^i\mathbf{n}$. Thus, the vectors $\mathbf{k}_i$ are the results of consequent actions over $\mathbf{n}$ of a linear operator with matrix A.

Let $B = \|b_{ij}\|$ be a matrix formed with the elements b_{ij} which are the cofactors of a_{ji} in the matrix A. Show that $\mathbf{P} = B\mathbf{n}$. To do this we compare, for instance, the first components of those vectors. The first component of $B\mathbf{n}$ is

$$b_{1j}\xi^j = \xi^1 \begin{vmatrix} \xi_2^2 & \cdots & \xi_{m+1}^2 \\ \cdots & \cdots & \cdots \\ \xi_2^{m+1} & \cdots & \xi_{m+1}^{m+1} \end{vmatrix} - \xi^2 \begin{vmatrix} \xi_2^1 & \cdots & \xi_{m+1}^1 \\ \cdots & \cdots & \cdots \\ \xi_2^{m+1} & \cdots & \xi_{m+1}^{m+1} \end{vmatrix} + \cdots$$

$$\cdots + (-1)^m \xi^{m+1} \begin{vmatrix} \xi_2^1 & \cdots & \xi_{m+1}^1 \\ \cdots & \cdots & \cdots \\ \xi_2^m & \cdots & \xi_{m+1}^m \end{vmatrix}$$

$$= \begin{vmatrix} \xi^1 & \xi_2^1 & \cdots & \xi_{m+1}^1 \\ \xi^2 & \xi_2^2 & \cdots & \xi_{m+1}^2 \\ \cdots & \cdots & \cdots & \cdots \\ \xi^{m+1} & \xi_2^{m+1} & \cdots & \xi_{m+1}^{m+1} \end{vmatrix}.$$

The first component of **P** has the form

$$(-1)^m(\mathbf{n}_2, \mathbf{n}_3, \ldots, \mathbf{n}_{m+1}, \mathbf{n}) = \left(\begin{vmatrix} \mathbf{e}_1 & \mathbf{e}_2 & \cdots & \mathbf{e}_{m+1} \\ \xi_2^1 & \xi_2^2 & \cdots & \xi_2^{m+1} \\ \cdots & \cdots & \cdots & \cdots \\ \xi_{m+1}^1 & \xi_{m+1}^2 & \cdots & \xi_{m+1}^{m+1} \end{vmatrix} . \mathbf{n}\right)$$

$$= \begin{vmatrix} \xi_1^1 & \xi_2^1 & \cdots & \xi_{m+1}^1 \\ \xi_1^2 & \xi_2^2 & \cdots & \xi_{m+1}^2 \\ \cdots & \cdots & \cdots & \cdots \\ \xi_1^{m+1} & \xi_2^{m+1} & \cdots & \xi_{m+1}^{m+1} \end{vmatrix}.$$

Thus, the first component of **P** coincides with the first one of $B\mathbf{n}$. The analogous result is true for all other components. Denote by $C = \|C_{ij}\|$ the matrix $-A$. In [75], p. 93 the definition of the adjoint matrix, denoted $B(\mu)$, of the matrix C is given. The element $b_{ij}(\mu)$ of the matrix $B(\mu)$ is a cofactor of $\mu\delta_{ij} - C_{ij}$ in the matrix $\|\mu\delta_{ij} - C_{ij}\|$. In a particular case, the matrix $B(0)$ is formed with cofactors in the matrix $-C$, i.e. with cofactors in A. Therefore, $B(0) = B$. In [75], p. 94 the expression for $B(\mu)$ in terms of the coefficients of the characteristic polynomial $\Delta(\mu) = |\mu\delta_{ij} - C_{ij}|$ and the powers of the matrix is presented. If we write the characteristic polynomial of C as

$$\Delta(\mu) = \mu^{m+1} - p_1\mu^m - \cdots - p_{m+1},$$

then for $\mu = 0$ we obtain

$$B(0) = C^m - p_1 C^{m-1} - \cdots - p_m E,$$

where E is a unit matrix.

Since $A = -C$,

$$B(0) = (-1)^m A^m - p_1(-1)^{m-1}A^{m-1} - \cdots + p_{m-1}A - p_m E.$$

Let us find p_i in terms of symmetric polynomials of principal curvatures of the second kind $\mu_1, \ldots, \mu_m$. They are the roots of the equation

$$|\mu\delta_{ij} + \xi_j^i| = |\mu\delta_{ij} - C_{ij}| = 0,$$

i.e. $\mu_1, \ldots, \mu_m$ are the eigenvalues of matrix C. Thus, we have the following expression for p_i:

$$p_1 = S_1,\ p_2 = -S_2,\ \ldots,\ p_i = (-1)^{i-1}S_i, \ldots.$$

Hence

$$B = B(0) = (-1)^m\{S_m E + S_{m-1}A + \cdots + A^m\}.$$

As $\mathbf{P} = B\mathbf{n}$,

$$\mathbf{P} = (-1)^m\{S_m\mathbf{n} + S_{m-1}A\mathbf{n} + \cdots + A^m\mathbf{n}\}.$$

Replacing the $A^i\mathbf{n}$ here with $\mathbf{k}_i$, we come to formula (10).

Using (9), we obtain the following *generalized Gauss–Bonnet formula for the case of a vector field in $(m+1)$-dimensional Euclidean space*:

$$(-1)^m \int_{F^m} (S_m \mathbf{n} + S_{m-1}\mathbf{k}_1 + \cdots + \mathbf{k}_m, \nu)\, dv = \omega_m \theta.$$

In the case when the field **n** coincides on F^m with the field of normals we come to the generalization of the Gauss–Bonnet formula to the hypersurface.

Observe some properties of the vector field **P**. Just as in Section 1.5, it is easy to state the *geometrical meaning of the field* **P** *streamlines: the field* **n** *is constant along them*. In the previous section we denoted these curves by M^1. From (9) it evidently follows that if $\mathbf{P} \equiv 0$ then the volume of the ψ image in S^m of any hypersurface F^m is zero, i.e. the dimension of the image is lower than m. Therefore, the vector **P** characterizes the curvature of **n** in dimension m. However, if $\mathbf{P} \equiv 0$ then the field **n** is not necessarily constant.

Since $(\mathbf{n}_1, \ldots, \mathbf{n}_{m+1}) = 0$, it is easy to check that $\operatorname{div} \mathbf{P} = 0$.

2.4 The Family of Parallel Hypersurfaces in Riemannian Space

Let us be given a hypersurface F in a Riemannian space R^n. Let $\mathbf{n}(x)$ be a regular field of normals of $F(x)$ defined in some domain $G \subset F$. We draw a geodesic of ambient space in the direction of $\mathbf{n}(x)$ through each point $x \in F$. We shall call the field of tangent vectors of geodesic lines the *geodesic field*. Denote it with the same letter **n**. Let M be a point in a distance s on the geodesic from $x \in F$ in the direction of $\mathbf{n}(x)$. If x ranges over all points in the domain G then the set of marked points M form, in general, some hypersurface. That hypersurface is called *geodesically parallel* or *parallel* simply with respect to the hypersurface F. Denote this surface by F_s. Considering various small fixed values of s, we obtain a family of geodesically parallel hypersurfaces filling some domain in the Riemannian space. The geodesic lines are orthogonal to those hypersurfaces at each point. If we take s to be large then it may happen that the geodesics from different points of F intersect each other and, as a consequence, either we have singular points in F_s or F_s degenerates into submanifolds of lower dimension, moreover into points.

We consider the following question: *how do the principal curvatures of geodesically parallel hypersurfaces change in moving along the geodesic which is orthogonal to the surfaces of the family*. A tangent direction τ on a hypersurface of a Riemannian space is called principal if the covariant derivative of **n** in the direction of τ is collinear with τ:

$$\nabla_\tau \mathbf{n} = -\lambda\tau.$$

The coefficient λ is called the *principal curvature of the hypersurface*. At each point of the hypersurface we have $n-1$ principal directions.

Theorem *Let τ be a principal direction on a hypersurface from the family of geodesically parallel hypersurfaces and λ the a corresponding principal curvature. Then* *the derivative of λ with respect to the arc length parameter of the geodesic orthogonal to the hypersurface has the following form*

$$\frac{d\lambda}{ds} = \lambda^2 + K(\tau, \mathbf{n}), \tag{1}$$

where $K(\tau, \mathbf{n})$ is the curvature of space with respect to the plane generated by τ and $\mathbf{n}$.

Let us be given a coordinate system in the space domain under consideration. As usual, we denote the vector contravariant components by superscripts and the covariant derivative by "comma". In these notations the equation for principal directions has the form

$$n^i{}_{,j}\tau^j = -\lambda\tau^i. \tag{2}$$

Differentiating both sides of (2) in the direction of $\mathbf{n}$, we obtain

$$n^i{}_{,jk}\tau^j n^k + n^i{}_{,j}\tau^j{}_{,k}n^k = -\frac{d\lambda}{ds}\tau^i - \lambda\tau^i{}_{,k}n^k. \tag{3}$$

Multiply both sides by τ. Since τ is a unit vector field, $(\tau_{,k}n^k, \tau) = 0$ and hence the right-hand side reduces itself to $-d\lambda/ds$. We have

$$\tau_i n^i{}_{,jk}\tau^j n^k + \tau_i n^i{}_{,j}\tau^j{}_{,k}n^k = -\frac{d\lambda}{ds}. \tag{4}$$

Consider the second term on the left-hand side of (4). The $\tau_{,k}n^k$ is orthonormal to τ, so that we have the following decomposition

$$\tau^j{}_{,k}n^k = \mu n^j + \sum_\alpha \nu_\alpha \tau_\alpha, \tag{5}$$

where τ_α stands for the principal direction different from τ and having λ_α as the corresponding principal curvature, μ and ν_α are some numbers. Let us find μ. To do this we multiply both sides of (5) by $\mathbf{n}$ scalarly:

$$n_j\tau^j{}_{,k}n^k = \mu.$$

Transform the expression on the left-hand side. Bring n_j under the sign of the covariant derivative in k and as a compensation subtract $\tau^j n_{j,k}n^k$. Then we get

$$(n_j\tau^j)_{,k}n^k - \tau^j n_{j,k}n^k = \mu.$$

Since $\mathbf{n}$ is orthogonal to τ and the field $\mathbf{n}$ is a geodesic one: $n_{j,k}n^k = 0$, as a consequence, $\mu = 0$. Using (5) with $\mu = 0$, write

$$\tau_i n^i{}_{,j}\tau^j{}_{,k}n^k = \tau_i n^i{}_{,j}\nu_\alpha\tau^j_\alpha = \nu_\alpha\lambda_\alpha\tau^i_\alpha\tau_i = 0. \tag{6}$$

Consider the first term on the left-hand side of (4):

$$\begin{aligned}\tau_i n^i{}_{,jk}\tau^j n^k &= \tau_i(n^i{}_{,kj} - R^i_{\cdot ljk}n^l)\tau^j n^k \\ &= (n^i{}_{,k}n^k)_{,j}\tau^j\tau_i - n^i{}_{,k}n^k{}_{,j}\tau^j\tau_i - K(\tau, \mathbf{n}),\end{aligned} \tag{7}$$

where $R^i_{\cdot ljk}$ is a curvature tensor of R^n.

The first term on the right-hand side of (7) is zero because **n** is a geodesic field. The second term is

$$-n^i{}_{;k}n^k{}_{,j}\tau^j\tau_i = n^i{}_{,k}\lambda\tau^k\tau_i = -\lambda^2. \tag{8}$$

So, after applying (6)–(8), equality (4) gives (1).

If we set $\lambda = -\frac{d\ln\eta}{ds}$ then (1) has the form of the Jacobi equation

$$\frac{d^2\eta}{ds^2} = -K(\tau,\mathbf{n})\eta.$$

This equation allows us to estimate the change in principal curvatures of geodesically parallel hypersurfaces along the geodesics orthogonal to the family under the given restrictions on the curvature of space.

2.5 The Constant Vector Fields and the Killing Fields

The field $\mathbf{n} = \{\xi^k\}$ in Riemannian space R^n is called *constant* if the covariant derivative of this field with respect to any direction is equal to zero:

$$\nabla_i\xi^k = 0.$$

Since $\mathbf{n}^2 = \xi_k\xi^k$, $\nabla_i\mathbf{n}^2 = 2\xi_k\nabla_i\xi^k = 0$. Therefore, the length of such a field is constant. If two constant vector fields are given then they make a constant angle. The curvature vector of the constant vector field streamline is $\xi^i\nabla_i\xi^k = 0$, i.e. the streamlines of the constant vector field are geodesic lines. Let **a** and **b** be orthogonal to **n**, i.e. $(\mathbf{a}.\mathbf{n}) = 0$, $(\mathbf{b}.\mathbf{n}) = 0$. Then

$$\begin{aligned}(\nabla_\mathbf{b}\mathbf{a} - \nabla_\mathbf{a}\mathbf{b}.\mathbf{n}) &= (\nabla_\mathbf{b}\mathbf{a}.\mathbf{n}) - (\nabla_\mathbf{a}\mathbf{b}.\mathbf{n})\\ &= \nabla_\mathbf{b}(\mathbf{a}.\mathbf{n}) - \nabla_\mathbf{a}(\mathbf{b}.\mathbf{n}) - (\mathbf{a}.\nabla_\mathbf{b}\mathbf{n}) + (\mathbf{b}.\nabla_\mathbf{a}\mathbf{n}) = 0.\end{aligned}$$

Hence, **n** is holonomic. Introduce the special system of coordinates $x^1,\ldots,x^n$ such that the surfaces $x^1 = \text{const}$ are orthogonal to **n**. As x^1 we take the distance from some fixed surface of $x^1 = \text{const}$ along the streamline. Then the first fundamental form of the Riemannian space has the following expression:

$$ds^2 = (dx^1)^2 + \sum_{\alpha,\beta=2}^{n} g_{\alpha\beta}\,dx^\alpha\,dx^\beta.$$

Since **n** is supposed to be constant, the functions $g_{\alpha\beta}$ do not depend on x^1. With respect to the coordinate system introduced the field **n** has the following components: $\xi^1 = 1$, $\xi^i = 0$, $i \geq 2$. Consider the derivative

$$\nabla_i\xi_k = \frac{\partial\xi_k}{\partial x^i} - \Gamma^j_{ki}\xi_j.$$

From this it follows that $\Gamma^1_{ik} = 0$. Since $g^{1\alpha} = 0$ for $\alpha \neq 1$,

$$\Gamma^1_{ik} = g^{11}\frac{1}{2}\left(\frac{\partial g_{1i}}{\partial x^k} + \frac{\partial g_{1k}}{\partial x^i} - \frac{\partial g_{ik}}{\partial x^1}\right) = -g^{11}\frac{1}{2}\frac{\partial g_{ik}}{\partial x^1} = 0.$$

Thus, if a constant vector field exists in a Riemannian space then the metric can be reduced to the following form:

$$ds^2 = (dx^1)^2 + \sum_{\alpha,\beta=2}^{n} g_{\alpha\beta}(x^2, \ldots, x^n)\, dx^\alpha\, dx^\beta$$

and vice versa.

From this it is easy to see that the *Riemannian space R^n which is not locally Euclidean admits no more then $n-2$ constant vector fields.*

Now we turn to the consideration of Killing vector fields. Let x^i be the local coordinates in Riemannian space R^n. The infinitesimal transformation of R^n is, by definition, the correspondence between the points of coordinates x^i and $\bar{x}^i$ given by

$$\bar{x}^i = x^i + \xi^i\,\delta t,$$

where ξ^i are components of some vector field ξ on R^n, δt is some fixed infinitesimal in R^n. The field ξ is, in general, of variable length. The infinitesimal shift $\delta x^i = \bar{x}^i - x^i$ has the form

$$\delta x^i = \xi^i\,\delta t.$$

We shall call ξ the *Killing field* if corresponding to ξ infinitesimal transformation the metric of space does not change up to the infinitesimals $(\delta t)^2$, i.e.

$$\delta\, ds^2 = 0$$

up to $(\delta t)^2$. In more detail, if ξ is the Killing field then

$$\delta\, ds^2 = g_{ij}(\bar{x}^\alpha)\, d\bar{x}^i\, d\bar{x}^j - g_{ij}(x^\alpha)\, dx^i\, dx^j$$

is the infinitesimal of a higher order than δt. We have

$$d\bar{x}^i = dx^i + \frac{\partial \xi^i}{\partial x^k}\, dx^k\,\delta t,$$

$$g_{ij}(\bar{x}^\alpha) = g_{ij}(x^\alpha) + \frac{\partial g_{ij}}{\partial x^\alpha}\,\delta x^\alpha = g_{ij}(x^\alpha) + \frac{\partial g_{ij}}{\partial x^\alpha}\,\xi^\alpha\,\delta t.$$

Up to infinitesimals $(\delta t)^2$, we have

$$d\bar{x}^i\, d\bar{x}^j = dx^i\, dx^j + \frac{\partial \xi^i}{\partial x^k}\, dx^k\, dx^j\,\delta t + \frac{\partial \xi^j}{\partial x^k}\, dx^k\, dx^i\,\delta t.$$

Hence

$$\delta\, ds^2 = \left(\frac{\partial g_{ij}}{\partial x^\alpha}\,\xi^\alpha\, dx^i\, dx^j + g_{ij}\frac{\partial \xi^i}{\partial x^k}\, dx^k\, dx^j + g_{ij}\frac{\partial \xi^j}{\partial x^k}\, dx^k\, dx^i\right)\delta t.$$

In the second term on the right we make the substitution $k \to i \to k$, and in the third $k \to j \to k$. We get

$$\delta\, ds^2 = \left(\frac{\partial g_{ij}}{\partial x^\alpha}\xi^\alpha + g_{kj}\frac{\partial \xi^k}{\partial x^i} + g_{ik}\frac{\partial \xi^k}{\partial x^j}\right) dx^i\, dx^j\, \delta t.$$

By the definition of a Killing field, $\delta\, ds^2$ must be the infinitesimal of higher order than δt in all choices of dx^i or dx^j. Therefore, the Killing field satisfies the following system of equations:

$$\frac{\partial g_{ij}}{\partial x^\alpha}\xi^\alpha + g_{kj}\frac{\partial \xi^k}{\partial x^i} + g_{ik}\frac{\partial \xi^k}{\partial x^j} = 0.$$

Since $\frac{\partial g_{ij}}{\partial x^\alpha} = \Gamma_{i\alpha,j} + \Gamma_{j\alpha,i}$, we are able to represent the latter equations as

$$g_{kj}\left(\frac{\partial \xi^k}{\partial x^i} + \Gamma^k_{i\alpha}\xi^\alpha\right) + g_{ik}\left(\frac{\partial \xi^k}{\partial x^j} + \Gamma^k_{j\alpha}\xi^\alpha\right) = 0.$$

We have the covariant derivatives in parentheses. Since the covariant derivative of a metric tensor is zero, *the equations for a Killing field can be expressed as*

$$\xi_{j.i} + \xi_{i.j} = 0. \tag{1}$$

These equations are called the *Killing equations*. The *divergence of Killing field is zero.* Indeed, contracting both sides of the Killing equations with g^{kj}, we obtain

$$\xi^k_{.i} = -g^{kj}\xi_{i.j}.$$

Further, setting $k = i$ and contracting in this index, we obtain

$$\operatorname{div}\xi = -\operatorname{div}\xi = 0.$$

If the length of a Killing field is constant then its streamlines are geodesics.

Show that the *second covariant derivatives of Killing field components ξ_i can be expressed in terms of these components and the Riemannian tensor.* Consider the Killing equations (1) for three pairs of indices

$$\xi_{i.j} + \xi_{j.i} = 0,$$
$$\xi_{j.k} + \xi_{k.j} = 0,$$
$$\xi_{k.i} + \xi_{i.k} = 0.$$

Covariantly differentiate the first equation in x^k, the second in x^i, the third in x^j. Add the first and second equations and subtract the third one. We get

$$\xi_{i.jk} - \xi_{i.kj} + \xi_{k.ji} - \xi_{k.ij} + \xi_{j.ki} + \xi_{j.ik} = 0.$$

Using the expression for the difference of second-order covariant derivatives of tensors, we get

$$(R^\alpha_{\cdot ijk} + R^\alpha_{\cdot jki} + R^\alpha_{\cdot kji})\xi_\alpha + 2\xi_{j.ik} = 0.$$

We can replace the sum of the two first components of the Riemannian tensor in parentheses with $R^{\alpha}_{\cdot kji}$. Hence

$$\xi_{j.ik} = -R^{\alpha}_{\cdot kji}\xi_{\alpha}. \quad (2)$$

Also, *there are the relations between the first-order derivatives of Killing field components and the original components*:

$$\xi_{\alpha}(R^{\alpha}_{\cdot lji.k} - R^{\alpha}_{\cdot kji.l}) + R^{\alpha}_{\cdot lji}\xi_{\alpha.k} - R^{\alpha}_{\cdot kji}\xi_{\alpha.l} - R^{\alpha}_{\cdot jkl}\xi_{\alpha.i} - R^{\alpha}_{\cdot ikl}\xi_{j.\alpha} = 0.$$

These relations arise as the compatibility conditions of (2).

2.6 On Symmetric Polynomials of Principal Curvatures of a Vector Field on Riemannian Space

The symmetric polynomials of principal curvatures of the second kind of a vector field with a constant length in Riemannian space have the same expressions as in the case of a vector field in Euclidean space if one replaces the usual derivatives with covariant derivatives:

$$S_k = (-1)^k \sum_{i_1<\dots<i_k} \begin{vmatrix} \xi^{i_1}_{.i_1} & \dots & \xi^{i_1}_{.i_k} \\ \dots & \dots & \dots \\ \xi^{i_k}_{.i_1} & \dots & \xi^{i_k}_{.i_k} \end{vmatrix}.$$

We are going to find the analogs of the divergent representations for S_2 and S_3. Let $\mathbf{k} = \{\xi^i_{.j}\xi^j\}$ be the field ξ streamline curvature vector. Consider the divergence of the field $-(S_1\xi + \mathbf{k})$:

$$\begin{aligned} \operatorname{div}(-S_1\xi - \mathbf{k}) &= (\xi^{i_1}_{.i_1}\xi^{i_2} - \xi^{i_2}_{.i_1}\xi^{i_1})_{.i_2} \\ &= (\xi^{i_1}_{.i_1}\xi^{i_2}_{.i_2} - \xi^{i_2}_{.i_1}\xi^{i_1}_{.i_2}) + (\xi^{i_1}_{.i_1 i_2}\xi^{i_2} - \xi^{i_2}_{.i_1 i_2}\xi^{i_1}) = 2S_2 + R_{\alpha\beta}\xi^{\alpha}\xi^{\beta}, \end{aligned}$$

where $R_{\alpha\beta}$ are the Ricci tensor components. The expression $-R_{\alpha\beta}\xi^{\alpha}\xi^{\beta}$ is the Ricci curvature for the direction ξ. It is equal to the sum of sectional curvatures of Riemannian space along the planes spanned on ξ and $n-1$ vectors which are mutually orthogonal and orthogonal to ξ. *Therefore, we have*

$$2S_2 = \operatorname{div}(-S_1\xi - \mathbf{k}) + \operatorname{Ric}(\xi).$$

In the case of a space of constant curvature K_0

$$2S_2 = \operatorname{div}(-S_1\xi - \mathbf{k}) + (n-1)K_0.$$

It is convenient further on to introduce a column

$$a = \begin{pmatrix} \xi^{i_1} \\ \xi^{i_2} \\ \xi^{i_3} \end{pmatrix}.$$

We shall denote the determinant of a matrix formed with this column and its derivatives by brackets. Then $6S_3 = [a_{.i_1} a_{.i_2} a_{.i_3}]$. Take out the sign of covariant differentiation and subtract the extra terms. We have

$$\begin{aligned} 18S_3 = {} & [aa_{.i_2} a_{.i_3}]_{.i_1} + [a_{.i_1} aa_{.i_3}]_{.i_2} + [a_{.i_1} a_{.i_2} a]_{.i_3} \\ & - [aa_{.i_2 i_1} a_{.i_3}] - [aa_{.i_2} a_{.i_3 i_1}] - [a_{.i_1 i_2} aa_{.i_3}] \\ & - [a_{.i_1} aa_{.i_3 i_2}] - [a_{.i_1 i_3} a_{.i_2} a] - [a_{.i_1} a_{.i_2 i_3} a]. \end{aligned}$$

Consider the sum of terms containing covariant derivatives of the second order. Gathering them in pairs, we get

$$-[a_{.i_1 i_2} - a_{.i_2 i_1} aa_{.i_3}] - [aa_{.i_2} a_{.i_3 i_1} - a_{.i_1 i_3}] - [a_{.i_1} aa_{.i_3 i_2} - a_{.i_2 i_3}].$$

Now we use the formula for the difference of second-order covariant derivatives of a tensor ξ^i. In detail the first of the determinants above is

$$[a_{.i_1 i_2} - a_{.i_2 i_1} aa_{.i_3}] = \begin{vmatrix} R^{\alpha i_1}_{\cdot\cdot i_1 i_2} \xi_\alpha & \xi^{i_1} & \xi^{i_1}_{.i_3} \\ R^{\alpha i_2}_{\cdot\cdot i_1 i_2} \xi_\alpha & \xi^{i_2} & \xi^{i_2}_{.i_3} \\ R^{\alpha i_3}_{\cdot\cdot i_1 i_2} \xi_\alpha & \xi^{i_3} & \xi^{i_3}_{.i_3} \end{vmatrix}$$

$$= -2R_{\alpha\beta} \xi^\alpha \xi^\beta S_1 + R^{\cdot i}_{\alpha\cdot\beta j} \xi^\alpha \xi^\beta \xi^j_{.i} - 2R_{\alpha\beta} \xi^\alpha k^\beta.$$

It is easy to check that the other two determinants containing second-order derivatives of a are equal to the first one. Now consider the determinant $[aa_{.i_2} a_{.i_3}]$ which stands under covariant differentiation in x_{i_1}. Expanding it with respect to the first column, we have

$$[aa_{.i_2} a_{.i_3}] = \xi^{i_1}(\xi^{i_2}_{.i_2} \xi^{i_3}_{.i_3} - \xi^{i_3}_{.i_2} \xi^{i_2}_{.i_3}) - \xi^{i_2}(\xi^{i_1}_{.i_2} \xi^{i_3}_{.i_3} - \xi^{i_3}_{.i_2} \xi^{i_1}_{.i_3}) + \xi^{i_3}(\xi^{i_1}_{.i_2} \xi^{i_2}_{.i_3} - \xi^{i_2}_{.i_2} \xi^{i_1}_{.i_3}).$$

Evidently, the expression in the first parentheses is equal to $2S_2$. Now gather the terms containing $\xi^{i_3}_{.i_3}$ and $\xi^{i_2}_{.i_2}$. The other two terms are equal to each other after summation over the index change. We obtain

$$[aa_{.i_2} a_{.i_3}] = 2(\xi^{i_1} S_2 + k^{i_1} S_1 + \xi^{i_1}_{.\alpha} \xi^\alpha_{.\beta} \xi^\beta).$$

The expression above is the i-th component of the vector **P** which we introduced for the case of a vector field in Euclidean space. Note that the expressions for the other two determinants, namely $[a_{.i_1} aa_{.i_3}]$ and $[a_{.i_1} a_{.i_2} a]$, can be obtained from the expressions presented by the substitution of i_1 by i_2 and i_3 respectively. Therefore, *we obtain the following formula for the third symmetric polynomial S_3 in the case of a vector field in Riemannian space:*

$$3S_3 = \operatorname{div} \mathbf{P} + R_{\alpha\beta} \xi^\alpha \xi^\beta S_1 + R_{\alpha\beta} \xi^\alpha k^\beta - R^{\cdot\cdot i}_{\alpha\cdot\beta j} \xi^\alpha \xi^\beta \xi^j_{.i}.$$

Consider the expression for the sum of terms with Riemannian and Ricci tensors in the case of space of constant curvature K_0. We have

$$-R_{\alpha\beta} \xi^\alpha \xi^\beta = (n-1)K_0, \quad R_{\alpha\beta} \xi^\alpha k^\beta = \frac{R}{n} g_{\alpha\beta} \xi^\alpha k^\beta = 0.$$

Introduce a coordinate system in such a way that at some fixed point $g_{\alpha\beta} = \delta_{\alpha\beta}$ and the coordinate curve x^1 is tangent to ξ. Then

$$R^{\cdot i}_{\alpha\cdot\beta j}\xi^\alpha \xi^\beta \xi^j_{,i} = -R_{1i1j}\xi^j_{,i} = K_0 S_1$$

So, *in the case of a space of constant curvature the third symmetric polynomial of the principal curvatures of the second kind of a vector field has the following expression*:

$$3S_3 = \operatorname{div}\mathbf{P} - (n-2)K_0 S_1.$$

Since S_1 is the divergence of a vector field $-\xi$, then S_3, in contrast to S_2, can be represented as a divergence of some vector field, namely of $\frac{1}{3}(\mathbf{P} + (n-2)K_0\xi)$.

2.7 The System of Pfaff Equations

Consider the expression of the form

$$\omega = a_1\,dx^1 + \cdots + a_n\,dx^n, \tag{1}$$

where $a_i = a_i(x^1, \ldots, x^n)$ are differentiable functions of coordinates $x^1, \ldots, x^n$. This expression is linear with respect to dx^i and is called *the Pfaff form*.

We shall suppose that in a coordinate change the set of functions a_i satisfies the transformation law of covariant tensor components. Namely, if a'_k are its components with respect to a new system of coordinates $y^1, \ldots, y^n$ then

$$a_i = a'_k \frac{\partial y^k}{\partial x^i}.$$

In this case the form expression is invariant with respect to the coordinate change:

$$\omega = a_i\,dx^i = a'_k \frac{\partial y^k}{\partial x^i} dx^i = a'_k\,dy^k.$$

Consider the two symbols of differentiation, namely δ and d. We shall adopt that $d\,\delta x^i = \delta\,dx^i$. The differentiation δ produces the differentials δx^i. We denote the Pfaff form with respect to δx^i by $\omega(\delta)$.

To each Pfaffian form ω we are able to put into correspondence the form of second order, namely $d\omega(\delta) - \delta\omega(d)$ which is called *the bilinear Frobenius form*.

In differentiating ω we use the functional character of a_i. We have

$$\delta\omega(d) = \delta(a_i\,dx^i) = \delta a_i\,dx^i + a_i\delta\,dx^i = \frac{\partial a_i}{\partial x^j}\,\delta x^j\,dx^i + a_i\delta\,dx^i,$$

$$d\omega(\delta) = \frac{\partial a_k}{\partial x^j}\,dx^j\delta x^k + a_k d\,\delta x^k.$$

From this we find the expression for the bilinear Frobenius form:

$$d\omega(\delta) - \delta\omega(d) = \frac{\partial a_k}{\partial x^j}\,dx^j\,\delta x^k - \frac{\partial a_k}{\partial x^j}\,\delta x^j\,dx^k = \frac{1}{2}\left(\frac{\partial a_k}{\partial x^j} - \frac{\partial a_j}{\partial x^k}\right)(dx^j\,\delta x^k - \delta x^j\,dx^k).$$

It is easy to check that the coefficients $\left(\frac{\partial a_k}{\partial x^i} - \frac{\partial a_i}{\partial x^k}\right)$ form a tensor. The equation $\omega = 0$ is called *the Pfaff equation*. Let us also define the bilinear form $\omega(d, \delta)$ which depends on both shifts $d\mathbf{r}$ and $\delta\mathbf{r}$:

$$\omega(d, \delta) = a_{ij}(dx^i\, \delta x^j - dx^j\, \delta x^i),$$

where a_{ij} form a tensor of second order, i.e.

$$a_{ij} = a'_{kl}\, \frac{\partial y^k}{\partial x^i}\, \frac{\partial y^l}{\partial x^j}.$$

The form of this expression is invariant in coordinate change:

$$\begin{aligned}\omega(d, \delta) &= a'_{kl}\, \frac{\partial y^k}{\partial x^i}\, \frac{\partial y^l}{\partial x^j}\, (dx^i\, \delta x^j - dx^j\, \delta x^i)\\ &= a'_{kl}(dy^k\, \delta y^l - dy^l\, \delta y^k).\end{aligned}$$

Therefore, if the form ω is zero with respect to some coordinate system for some shifts $d\mathbf{r}$ and $\delta\mathbf{r}$ then it is zero with respect to any other coordinate system for these shifts.

Let us be given the system of Pfaff equations in a domain $G \subset E^n$

$$\begin{cases}\omega_1 = a_{1i}\, dx^i = 0, & \mathrm{i} = 1, \ldots, \mathrm{n},\\ \ldots & \\ \omega_m = a_{mi}\, dx^i = 0, & \mathrm{m} < \mathrm{n}\,.\end{cases} \tag{2}$$

We say that (2) *is not degenerate at each point a if the rank of the matrix* $\|a_{ij}\|$ *is equal to m*. This definition can be interpreted geometrically. Consider the system of m vectors $\mathbf{A}_1 = \{a_{1i}\}, \ldots, \mathbf{A}_m = \{a_{mi}\}$ in a Euclidean space. The condition of non-degeneracy means that $\mathbf{A}_1, \ldots, \mathbf{A}_m$ are linearly independent. We denote the space they span by T^m (see Fig. 21). The surface V_{n-m} of dimension $n - m$ in Euclidean space E^n is called the integral surface if any shift $\{dx^i\}$ in tangent space of V_{n-m} satisfies system (2). This surface is orthogonal to T^m at each point. In this case, i.e. when V_{n-m} exists, we say that *the Pfaff system is totally integrable or holonomic*.

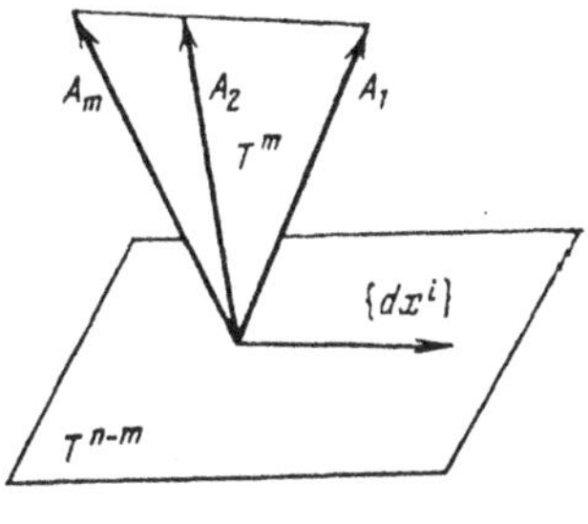

FIGURE 21

The condition, under which the system (2) is totally integrable, gives the following theorem.

Frobenius theorem *The Pfaff system of equations $\omega_j = 0$, $j = 1, \ldots, m$ is totally integrable in a domain $G \subset E^n$ if and only if for all ω_j the bilinear Frobenius forms are zero by virtue of the system $\omega_j = 0$.*

We cite the proof from [26]. The conversion to zero of bilinear Frobenius forms by virtue of the system $\omega_j = 0$ means that these bilinear forms are zero for all shifts $\{dx^i\}$ and $\{\delta x^i\}$ which satisfy the system (2).

Prove, firstly, the necessity. Suppose that the system $\omega_j = 0$ is totally integrable.

Let T^{n-m} be a space perpendicular to T^m. Consider the linear mapping M of E^n onto T^m which is the orthogonal projection onto T^m at each point $P \in G$. This mapping maps all the vectors of T^{n-m} into zero and is identical on T^m. By the hypothesis, the integral surface V_{n-m} of (2) exists. Let

$$\mathbf{r} = \mathbf{r}(u_1, \ldots, u_{n-m}).$$

be its parametric representation.

The tangent vectors $\mathbf{r}_{u_i}$ of this surface are in T^{n-m}. Hence,

$$M\mathbf{r}_{u_i} = 0, \quad i = 1, \ldots, n-m. \tag{3}$$

The linear mapping M with respect to the fixed coordinate system in E^n, at a fixed point P is determined by the matrix of order n. The elements of this matrix are the continuously differentiable functions of P, i.e. of parameters u_i. The derivative of M means, in our considerations, the derivative of the corresponding matrix.

Differentiate (3) in u_j. We get

$$M\mathbf{r}_{u_i u_j} + M_{u_j}\mathbf{r}_{u_i} = 0. \tag{4}$$

Interchanging the roles of i and j and subtracting the result from (4), we find

$$M_{u_j}\mathbf{r}_{u_i} - M_{u_i}\mathbf{r}_{u_j} = 0.$$

Multiply this by δu_i and du_j. Then sum over i and j from 1 to $n-m$. Then we obtain

$$(M_{u_i}\,\delta u_j)(\mathbf{r}_{u_i}\,du_i) - (M_{u_i}\,du_i)(\mathbf{r}_{u_j}\,\delta u_j) = 0. \tag{5}$$

We can express the first differentials of M and $\mathbf{r}$ in terms of the coordinate x_i in E^n:

$$M_{u_j}\,\delta u_j = M_{x^i}\,\delta x^i, \quad M_{u_i}\,du_i = M_{x^i}\,dx^i,$$
$$\mathbf{r}_{u_i}\,du_i = \mathbf{r}_{x^j}\,dx^j = d\mathbf{r}, \quad \mathbf{r}_{u_j}\,\delta u_j = \mathbf{r}_{x^k}\,\delta x^k,$$

where $d\mathbf{r}$ and $\delta\mathbf{r}$ are tangent to V_{n-m}, i.e. they are in T^{n-m}. To simplify notations we set

$$M_{x^i}\,\delta x^i = M_{\delta\mathbf{r}}, \quad M_{x^i}\,dx^i = M_{d\mathbf{r}}.$$

We rewrite the equation (5) in a brief form

$$M_{\delta\mathbf{r}}\,d\mathbf{r} - M_{d\mathbf{r}}\,\delta\mathbf{r} = 0, \tag{6}$$

where $d\mathbf{r}$, $\delta\mathbf{r} \in T^{n-m}$. Express condition (6) in terms of the coefficients of a given Pfaffian system $\omega_j = 0$. To do this, we find the matrix of M. Suppose that M takes the vector $\{dx^k\}$ into the vector $\{d\tilde{x}^j\}$ by

$$d\tilde{x}^j = \sum_k m_{jk}\,dx^k,$$

where $\|m_{jk}\|$ is the matrix of M. Since M takes any vector $\{dx^i\}$ which satisfies the Pfaff system (6) to zero, $\sum_k m_{jk}\,dx^k$ is a linear combination of systems of Pfaff equations

$$m_{jk}\,dx^k = \lambda_{js}a_{sk}\,dx^k, \quad j = 1, \dots, n,$$

where λ_{js} are some indefinite multipliers. The latter expression holds for all shifts $\{dx^k\}$. Therefore, $m_{jk} = \lambda_{js}\,a_{sk}$. Since M takes all of the $\mathbf{A}_s = \{a_{s1}, \dots, a_{sn}\}$ into themselves,

$$\lambda_{js}\,a_{sk}\,a_{lk} = a_{lj}.$$

Here the first subscript in a_{lj} is a number of $\mathbf{A}_l$ and the second is the number of these vector components with respect to E^n. The matrix $\|a_{lj}\|$ is not square, but it may be completed to a square matrix in such a way that the determinant will be non-zero. This corresponds to the possibility of completing m independent vectors $\mathbf{A}_l$ to a basis in E^n. We denote the complemented matrix by A. Then we may take the components of the inverse matrix A^{-1} as the λ_{ji}. Consider (6). We have

$$M_{\delta\mathbf{r}} = \frac{\partial M}{\partial x^\alpha}\,\delta x^\alpha.$$

The matrix $\partial M/\partial x^\alpha$ consists of elements $\partial m_{jk}/\partial x^\alpha$. Therefore, we have

$$M_{\delta\mathbf{r}} = \left\|\frac{\partial m_{jk}}{\partial x^\alpha}\,\delta x^\alpha\right\|.$$

The components of $M_{\delta\mathbf{r}}\,d\mathbf{r} - M_{d\mathbf{r}}\,\delta\mathbf{r}$ have the form

$$\begin{aligned}\frac{\partial m_{jk}}{\partial x^\alpha}\,\delta x^\alpha\,dx^k - \frac{\partial m_{jk}}{\partial x^\alpha}\,dx^\alpha\,\delta x^k &= \frac{\partial \lambda_{js}\,a_{sk}}{\partial x^\alpha}\,\delta x^\alpha\,dx^k - \frac{\partial \lambda_{js}\,a_{sk}}{\partial x^\alpha}\,dx^\alpha\,\delta x^k \\ &= \lambda_{js}\left(\frac{\partial a_{sk}}{\partial x^\alpha} - \frac{\partial a_{s\alpha}}{\partial x^k}\right)dx^k\,\delta x^\alpha \\ &\quad + \frac{\partial \lambda_{js}}{\partial x^\alpha}\,\delta x^\alpha(a_{sk}\,dx^k) - \frac{\partial \lambda_{js}}{\partial x^\alpha}\,dx^\alpha(a_{sk}\,\delta x^k).\end{aligned}$$

By virtue of (6), the latter expression is zero.

The shifts $d\mathbf{r}$, $\delta\mathbf{r} \in T^{n-m}$. Therefore, the last two terms are zero. Since the determinant $|\lambda_{js}| \neq 0$, the system

$$\lambda_{js}(\delta\omega_s(d) - d\omega_s(\delta)) = 0$$

has trivial solutions only, i.e. for all s the linear Frobenious forms are zero $\delta\omega_s(d) - d\omega_s(\delta) = 0$ by virtue of the system $\omega_s(d) = \omega_s(\delta) = 0$. Thus, the necessary condition is proved.

Now prove that this condition is sufficient, also, for the system $\omega_j = 0$ to be holonomic. We shall construct some surface and prove that this surface is tangent to T^{n-m} at each of its points. Consider at $P_0 \in G$ a unit vector $\tau \in T^{n-m}(P_0)$ with the initial point at P_0. Its end-point lies in a unit sphere S which is centered at P_0 and located in $T^{n-m}(P_0)$. The dimension of this sphere is $n - m - 1$. Introduce the coordinates $u_1, \ldots, u_{n-m-1}$ in this sphere. Then τ can be considered as a vector function of these coordinates:

$$\tau = \tau(u_1, \ldots, u_{n-m-1}).$$

Span the $(m+1)$-dimensional plane $T^{m+1}(P_0)$ on τ and $T^m(P_0)$. Let P be an arbitrary point in $T^{m+1}(P_0)$. At each point P the corresponding plane $T^m(P)$ is defined. Project $T^m(P)$ into $T^{m+1}(P_0)$. In a sufficiently small neighborhood of P_0 the projection is a one-to-one mapping and the result of projection is of dimension m. Denote it by $\bar{T}^m(P)$ (see Fig. 22). At each point $P \in T^{m+1}(P_0)$ we have a vector field orthogonal to $\bar{T}^m(P)$. Draw the streamline of this field through P_0 in the direction of τ. This curve is orthogonal to $T^m(P)$ because it is orthogonal to the projection of $T^m(P)$ into $T^{m+1}(P_0)$. If we take the various τ at P_0 and draw the curves γ as above then we obtain some $(n-m)$-dimensional surface of the position vector

$$\mathbf{r} = \mathbf{r}(u_1, \ldots, u_{n-m-1}, \sigma),$$

where σ is the arc length parameter in γ. By construction, $\mathbf{r}_\sigma$ being tangent to γ is orthogonal to $T^m(P)$. Hence,

$$M\mathbf{r}_\sigma = 0. \tag{7}$$

Show that

$$M\mathbf{r}_{u_i} = 0, \quad i = 1, \ldots, n - m - 1. \tag{8}$$

This means that all tangent vectors of the constructed surface lie in $T^{n-m}(P)$ for every P, i.e. the surface is the integral one of the Pfaff system (2) and, as a consequence of the definition, the system is holonomic. Introduce a vector $\mathbf{a} = M\mathbf{r}_{u_i}$ and find a linear differential equation this vector satisfies. We have

$$\mathbf{a}_\sigma = M\mathbf{r}_{u_i\sigma} + \frac{\partial M}{\partial \sigma}\mathbf{r}_{u_i} = M\mathbf{r}_{u_i\sigma} + \frac{\partial M}{\partial x^j}\frac{\partial x_j}{\partial \sigma}\mathbf{r}_{u_i} = M\mathbf{r}_{u_i\sigma} + M_{\mathbf{r}_\sigma}\mathbf{r}_{u_i}. \tag{9}$$

If we differentiate (7) in u_i, we get

$$M\mathbf{r}_{\sigma u_i} + M_{u_i}\mathbf{r}_\sigma = M\mathbf{r}_{\sigma u_i} + \frac{\partial M}{\partial x^j}\frac{\partial x_j}{\partial u_i}\mathbf{r}_\sigma = M\mathbf{r}_{\sigma u_i} + M_{\mathbf{r}_{u_i}}\mathbf{r}_\sigma = 0. \tag{10}$$

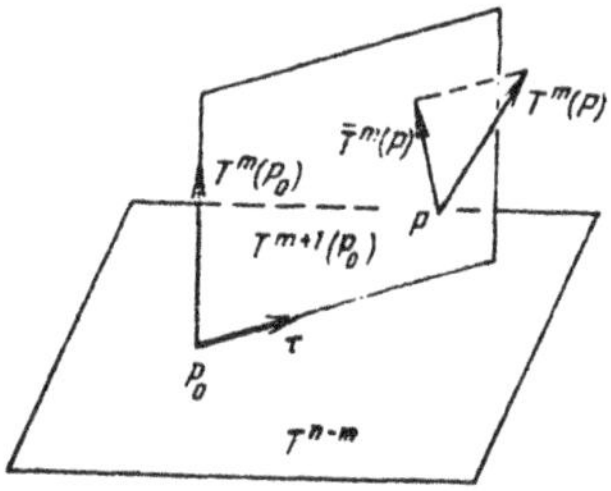

FIGURE 22

Subtracting (10) from (9), we obtain

$$\mathbf{a}_\sigma = M_{\mathbf{r}_\sigma}\mathbf{r}_{u_i} - M_{\mathbf{r}_{u_i}}\mathbf{r}_\sigma. \tag{11}$$

By the hypothesis, all the bilinear Frobenius forms of ω_j are zeroes by virtue of the system (2). Therefore, the differentials $d\mathbf{r}, \delta\mathbf{r} \in T^{n-m}$ satisfy (6). If it were known that $\mathbf{r}_{u_i} \in T^{n-m}$ then we could apply (6). Consider the decomposition of $\mathbf{r}_{u_i}$,

$$\mathbf{r}_{u_i} = \mathbf{a} + \mathbf{b},$$

where $\mathbf{a} = M\mathbf{r}_{u_i} \in T^m(P)$ and $\mathbf{b} \in T^{n-m}(P)$. Also, we can write

$$M_{\mathbf{r}_{u_i}} = M_\mathbf{a} + M_\mathbf{b}.$$

Then (11) can be rewritten by virtue of (6) as

$$\mathbf{a}_\sigma = M_{\mathbf{r}_\sigma}(\mathbf{a} + \mathbf{b}) - (M_\mathbf{a} + M_\mathbf{b})\mathbf{r}_\sigma = M_{\mathbf{r}_\sigma}\mathbf{a} - M_\mathbf{a}\mathbf{r}_\sigma. \tag{12}$$

So, the vector $\mathbf{a}$ satisfies some linear differential equation. At P_0, where $\sigma = 0$, we have $\mathbf{r}_{u_i} = \mathbf{r}_{u_i}(u_1, \dots u_{n-m-1}, 0)$ because the initial point P_0 is the same for all γ starting from P_0. Therefore, $\mathbf{a}(0) = M\mathbf{r}_{u_i}(u_1, \dots u_{n-m-1}, 0) = 0$. Hence, (12) implies $\mathbf{a}(\sigma) \equiv 0$, i.e. at an arbitrary point P of the constructed surface we have $\mathbf{r}_{u_i} \in T^{n-m}(P)$. The theorem is proved.

Consider the particular case, when *the system (2) consists of a single equation*

$$\omega = \sum_{i=1}^{n} \xi_i\, dx^i = 0.$$

We may assume that in a domain $G \subset E^n$ a unit vector field $\mathbf{n} = \{\xi_i\}$ is defined. The field $\mathbf{n}$ is holonomic if the family of hypersurfaces having this field as a normal one exists. By the theorem just proved, this occurs if and only if the bilinear Frobenius form $d\omega(\delta) - \delta\omega(d)$ of the form ω is zero for the shifts $d\mathbf{r}$ and $\delta\mathbf{r}$ which are orthogonal to $\mathbf{n}$. In terms of the components of $\mathbf{n}$, $d\mathbf{r}$ and $\delta\mathbf{r}$ this condition can be expressed as

$$\sum_{\alpha,\beta}^{n} \left(\frac{\partial \xi_\alpha}{\partial x^\beta} - \frac{\partial \xi_\beta}{\partial x^\alpha}\right)(dx^\alpha\, \delta x^\beta - dx^\beta\, \delta x^\alpha) = 0. \tag{13}$$

Choose the system of coordinates in E^n in such a way that at a fixed point P the components of the vector $\mathbf{n}$ are $(0, 0, \dots, 1)$. Then for every admissable shifts $d\mathbf{r}$ and $\delta\mathbf{r}$ one must have $dx^n = 0$ and $\delta x^n = 0$. Hence, in (13) the summation over α and β goes from 1 to $n-1$. If we take the shifts $d\mathbf{r} = (0, \dots, \overset{\alpha}{1}, 0, \dots, 0)$ and $\delta\mathbf{r} = (0, \dots, \overset{\beta}{1}, \dots, 0)$, where the unit takes the α-th place in $d\mathbf{r}$ and the β-th place in $\delta\mathbf{r}$, then we obtain the system of $(n-1)(n-2)/2$ equations:

$$\frac{\partial \xi_\alpha}{\partial x^\beta} - \frac{\partial \xi_\beta}{\partial x^\alpha} = 0, \quad 1 \le \alpha, \quad \beta \le n-1. \tag{14}$$

Denote the derivative $\partial/\partial x^\alpha$ by ∂_α. *Consider the set of the following expressions with respect to an arbitrary system of coordinates in* E^n:

$$\Omega_{\alpha\beta\gamma} = \xi_{[\alpha}\partial_\beta\xi_{\gamma]},$$

where the brackets which include the indices α, β, γ mean the sum of various terms which can be obtained from $\xi_\alpha\, \partial_\beta\, \xi_\gamma$ with permutations of indices α, β, γ; also, the corresponding expression is included in this sum with a "+" sign if the permutation is even and a "−" sign if the permutation is odd. Consider the way $\Omega_{\alpha\beta\gamma}$ changes under the coordinate change in space. Let $x^\alpha = x^\alpha(y^i)$. Then the components ξ'_i of $\mathbf{n}$ with respect to the system y^i are related to its components with respect to the system x^α by the tensor rule

$$\xi'_i = \xi_\alpha \frac{\partial x^\alpha}{\partial y^i}.$$

Let us form the expression analogous to $\Omega_{\alpha\beta\gamma}$ with respect to the system y^i:

$$\begin{aligned}\Omega'_{ijk} &= \xi'_{[i}\, \partial'_j\, \xi'_{k]} = \xi_\alpha \frac{\partial x^\alpha}{\partial y^{[i}} \frac{\partial}{\partial x^\gamma}\left(\xi_\beta \frac{\partial x^\beta}{\partial y^k}\right) \frac{\partial x^\gamma}{\partial y^{j]}} \\ &= \xi_\alpha\, \partial_\gamma\, \xi_\beta \frac{\partial x^\alpha}{\partial y^{[i}} \frac{\partial x^\beta}{\partial y^k} \frac{\partial x^\gamma}{\partial y^{j]}} + \xi_\alpha\, \xi_\beta \frac{\partial x^\alpha}{\partial y^{[i}} \frac{\partial^2 x^\beta}{\partial y^k \partial y^{j]}}.\end{aligned}$$

The last term on the right is zero due to the equality to each other of the mixed derivatives of the functions $x^\beta = x^\beta(y)$ and the convention given above for the brackets. The first term can be transformed by relabelling the summation indices as

$$\Omega'_{ijk} = \xi_{[\alpha}\, \partial_\gamma\, \xi_{\beta]} \frac{\partial x^\alpha}{\partial y^i} \frac{\partial x^\beta}{\partial y^k} \frac{\partial x^\gamma}{\partial y^j} = \Omega_{\alpha\beta\gamma} \frac{\partial x^\alpha}{\partial y^i} \frac{\partial x^\gamma}{\partial y^j} \frac{\partial x^\beta}{\partial y^k}.$$

So, the quantities $\Omega_{\alpha\beta\gamma}$ *satisfy the tensor rule of transformation. If the field is holonomic then all components* $\Omega_{\alpha\beta\gamma} = 0$ *at the point* P *with respect to the system of coordinates chosen above*. Observe that if a tensor $\Omega_{\alpha\beta\gamma}$ and bilinear Frobenius form are zero with respect to a given system of coordinates then they are zero with respect to any other system of coordinates. Therefore, the condition $\Omega_{\alpha\beta\gamma} = 0$ is necessary and sufficient for the field $\mathbf{n}$ to be holonomic. In the particular case, when the field $\mathbf{n}$ is in a domain $G \subset E^3$, that condition gives $(\mathbf{n}, \operatorname{cunl} \mathbf{n}) = 0$. The tensor $\Omega_{\alpha\beta\gamma}$ is called the *object of non-holonomicity* of the field $\mathbf{n}$ in E^n.

2.8 An Example from the Mechanics of Non-Holonomic Constraints

In mechanics, the restrictions posed on the motion of the body can often be expressed in the form of a system of differential equations. This system may either be integrable or not. As a typical example, we consider the equations of constraints which arise in a disk D rolling over the xy-plane. Suppose that the disk is always perpendicular to the xy-plane and the disk plane intersects the xy-plane in a straight line which makes an angle φ with the x-axis.

Set O to be a center of D, R is a radius of D, A is a point of contact of D to the xy-plane, B is a fixed point on the boundary circle of D which was the point of contact at the initial instant.

Denote the angle between OA and OB by ψ (see Fig. 23). As the rolling of the disk is assumed to be without sliding, the length of the curve γ in the xy-plane is equal to the length of the circle arc AB. In an infinitesimal motion of A this arc of length $R\,d\psi$ is the infinitesimal length of γ, namely, $ds = \sqrt{dx^2 + dy^2}$. Hence, we have

$$\begin{aligned} dx &= R\cos\varphi\, d\psi, \\ dy &= R\sin\varphi\, d\psi. \end{aligned}$$

Consider the four-dimensional space E^4 of the parameters x, y, φ, ψ and the system of Pfaff equations

$$\begin{aligned} \omega_1 &= dx - R\cos\varphi\, d\psi = 0, \\ \omega_2 &= dy - R\sin\varphi\, d\psi = 0. \end{aligned} \tag{1}$$

The bilinear Frobenius forms are

$$\begin{aligned} d\omega_1(\delta) - \delta\omega_1(d) &= R\sin\varphi(d\varphi\,\delta\psi - \delta\varphi\, d\psi), \\ d\omega_2(\delta) - \delta\omega_2(d) &= -R\cos\varphi(d\varphi\,\delta\psi - \delta\varphi\, d\psi). \end{aligned} \tag{2}$$

The admissible shifts $(dx, dy, d\varphi, d\psi)$ and $(\delta x, \delta y, \delta\varphi, \delta\psi)$ satisfy the system (1). Therefore, they can be written as $(R\cos\varphi\, d\psi, R\sin\varphi\, d\psi, d\varphi, d\psi)$ and $(R\cos\varphi\,\delta\psi, R\sin\varphi\,\delta\psi, \delta\varphi, \delta\psi)$. Hence, the differentials $d\varphi, d\psi, \delta\varphi, \delta\psi$ are arbitrary in admissible

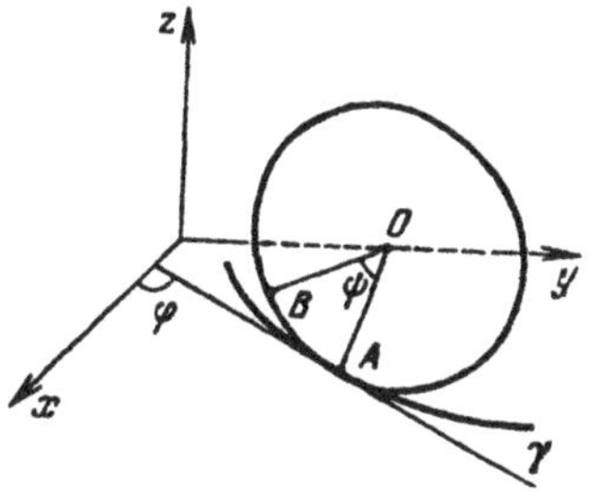

FIGURE 23

shifts and can be taken in such a way that $d\psi\,\delta\varphi - d\varphi\,\delta\psi \neq 0$. It follows from (2) that

$$\sum_{i=1}^{2}[\delta\omega_i(d) - d\omega_i(\delta)]^2 = R^2(d\varphi\,\delta\psi - \delta\varphi\,d\psi)^2 \neq 0.$$

Therefore, the system of equations of constraints for the disk rolling over the plane is non-holonomic.

2.9 The Exterior Differential Forms

Let $x^1, \ldots, x^n$ be curvilinear coordinates in a domain D of a Riemannian space V^n. The structure of exterior differential forms can be expressed in terms of the coordinate differentials dx^i. Introduce terms of the form

$$dx^{i_1} \wedge dx^{i_2} \wedge \ldots \wedge dx^{i_p}.$$

It is possible to multiply them by numbers and to add them to each other. Also, they possess a skew-symmetry property, i.e. if $j_1, \ldots, j_p$ is a permutation of numbers $i_1, \ldots, i_p$ then

$$dx^{i_1} \wedge dx^{i_2} \wedge \ldots \wedge dx^{i_p} = \varepsilon^{i_1 \ldots i_p}_{j_1 \ldots j_p}\, dx^{j_1} \wedge dx^{j_2} \wedge \ldots \wedge dx^{j_p},$$

where the generalized Kronecker symbol is

$$\varepsilon^{i_1 \ldots i_p}_{j_1 \ldots j_p} = \begin{cases} 1 & \text{if the permutation is even,} \\ -1 & \text{if the permutation is odd.} \end{cases}$$

The exterior differential form of order p with respect to local coordinates is the following sum:

$$\omega = \sum_{i_1 < \ldots < i_p} a_{i_1 \ldots i_p}\, dx^{i_1} \wedge \ldots \wedge dx^{i_p}$$

or

$$\omega = \frac{1}{p!}\sum_{i_1, \ldots, i_p} a_{i_1 \ldots i_p}\, dx^{i_1} \wedge \ldots \wedge dx^{i_p},$$

where the summation is taken over the indices $i_1 \ldots i_p$ which takes values from 1 to n independently and the form coefficients $a_{i_1 \ldots i_p}$ satisfy the skew-symmetry condition

$$a_{j_1 \ldots j_p} = \pm a_{i_1 \ldots i_p},$$

where "+" is assigned if the permutation $\begin{pmatrix} j_1 \ldots j_p \\ i_1 \ldots i_p \end{pmatrix}$ is even and "−" if the permutation is odd.

Suppose that in D we have other curvilinear coordinates $u^1, \ldots, u^n$ such that

$$x^i = x^i(u^1, \ldots, u^n).$$

Consider the coefficients of ω with respect to the system of coordinates u^{α}. We have

$$\omega = \frac{1}{p!}\sum_{i_1,\ldots,i_p} a_{i_1\ldots i_p}\frac{\partial x^{i_1}}{\partial u^{\alpha_1}}\cdots\frac{\partial x^{i_p}}{\partial u^{\alpha_p}}\,du_1^{\alpha}\wedge\ldots\wedge du^{\alpha_p}.$$

Therefore, the coefficients $a'_{\alpha_1\ldots\alpha_p}$ of ω with respect to a new system of coordinates have the form

$$a'_{\alpha_1\ldots\alpha_p} = a_{i_1\ldots i_p}\frac{\partial x^{i_1}}{\partial u^{\alpha_1}}\cdots\frac{\partial x^{i_p}}{\partial u^{\alpha_p}},$$

i.e. the quantities $a_{i_1\ldots i_p}$ form the covariant components of a tensor of order p.

As an example, we present the forms ω_i of order i in three-dimensional space

$$\begin{aligned}\alpha &= A\,dx^1 + B\,dx^2 + C\,dx^3,\\ \beta &= P\,dx^2\wedge dx^3 + Q\,dx^3\wedge dx^1 + R\,dx^1\wedge dx^2,\\ \gamma &= S\,dx^1\wedge dx^2\wedge dx^3,\end{aligned} \tag{1}$$

where $A,\ldots S$ are functions of x^1, x^2, x^3.

For two given exterior forms their exterior product is defined. Let ω_1 and ω_2 be the exterior forms of order p and q respectively:

$$\begin{aligned}\omega_1 &= \frac{1}{p!}\sum_{i_1,\ldots,i_p} a_{i_1\ldots i_p}\,dx^{i_1}\wedge\ldots\wedge dx^{i_p},\\ \omega_2 &= \frac{1}{q!}\sum_{j_1,\ldots,j_p} b_{j_1\ldots j_q}\,dx^{j_1}\wedge\ldots\wedge dx^{j_q}.\end{aligned} \tag{2}$$

The exterior product $\omega_1\wedge\omega_2$ of the forms ω_1 and ω_2 is the following form of order $p+q$:

$$\omega_1\wedge\omega_2 = \frac{1}{p!q!}\sum_{\substack{i_1\ldots i_p\\ i_1\ldots i_p}} a_{i_1\ldots i_p} b_{j_1\ldots j_q}\,dx^{i_1}\wedge\ldots\wedge dx^{i_p}\wedge dx^{j_1}\wedge\ldots\wedge dx^{j_q}. \tag{3}$$

For example, the exterior product of α and β from (1) is the form of order three, namely

$$\alpha\wedge\beta = (AP + BQ + CR)\,dx^1\wedge dx^2\wedge dx^3.$$

If we transpose the cofactors then the exterior product either stays the same or only changes the sign. Indeed, $\omega_2\wedge\omega_1$ can be obtained from $\omega_1\wedge\omega_2$ by transpositions of the differentials dx^{i_α} and dx^{j_β}. The number of those transpositions is pq. Hence,

$$\omega_1\wedge\omega_2 = (-1)^{pq}\omega_2\wedge\omega_1.$$

Let us define the exterior differential $d\omega$ of the exterior form ω. Denote the linear form by $da_{i_1\ldots i_p}$, namely the differential of the function $a_{i_1\ldots i_p}$. Then *the* *exterior differential of the form ω of order p is the following exterior form of order $p+1$:*

$$d\omega = \frac{1}{p!}\sum_{i_1\dots i_p} da_{i_1\dots i_p} \wedge dx^{i_1} \wedge \dots \wedge dx^{i_p}. \tag{4}$$

The following rule of differentiation holds for the exterior product the forms ω_1 and ω_2:

$$d(\omega_1 \wedge \omega_2) = d\omega_1 \wedge \omega_2 + (-1)^p \omega_1 \wedge d\omega_2. \tag{5}$$

Indeed, taking into account (3), we have

$$d\omega_1 \wedge \omega_2 = \frac{1}{p!q!}\sum_{\substack{i_1\dots i_p\\ j_1\dots j_p}}\{(da_{i_1\dots i_p})b_{j_1\dots j_q}\, dx^{i_1}\wedge\dots\wedge dx^{j_q} + a_{i_1\dots i_p} db_{j_1\dots j_q}\wedge dx^{i_1}\wedge\dots\wedge dx^{j_q}\}.$$

The first term is $d\omega_1 \wedge \omega_2$. In the second term we make p transpositions of the differential $db_{i_1\dots i_p}$ with the differentials $dx^{i_1},\dots,dx^{i_p}$. Then this term gives the form for the exterior product $\omega_1 \wedge (-1)^p d\omega_2$. Formula (5) is proved.

The following lemma plays an important role.

Poincaré lemma The second exterior differential of a differential form is identically zero.

$$d\,d\omega = 0.$$

Consider the expression for the exterior differential

$$d\omega = \frac{1}{p!}\sum_{\alpha.i_1,\dots,i_p} \frac{\partial a_{i_1\dots i_p}}{\partial x^\alpha}\, dx^\alpha \wedge \dots \wedge dx^{j_p}.$$

The coefficients of this form are the functions $\frac{\partial a_{i_1\dots i_p}}{\partial x^\alpha}$. Hence

$$\begin{aligned} d^2\omega &= \frac{1}{(p+1)!}\sum_{\alpha,i_1,\dots,i_p} d\frac{\partial a_{i_1\dots i_p}}{\partial x^\alpha}\wedge dx^\alpha \wedge dx^{i_1}\wedge\dots\wedge dx^{i_p} \\ &= \frac{1}{(p+1)!}\sum_{\alpha,\beta,i_1,\dots,i_p} \frac{\partial^2 a_{i_1\dots i_p}}{\partial x^\alpha x^\beta}\, dx^\beta \wedge dx^\alpha \wedge dx^{i_1}\wedge\dots\wedge dx^{i_p}. \end{aligned} \tag{6}$$

Since $dx^\beta \wedge dx^\alpha = -dx^\alpha \wedge dx^\beta$ and the second mixed derivatives of $a_{i_1\dots i_p}$ are equal to each other, the sum on the right of (6) contains the term of opposite sign for each term. Therefore, this sum is zero.

The inverse statement is also true: *if the exterior differential of some form ω is zero, i.e. $d\omega = 0$, then locally, in a sufficiently small neighborhood, the form ω is the exterior differential of some other form.*

In other words, *if $d\omega = 0$ then in a sufficiently small neighborhood there is a form ω_1 such that $\omega = d\omega_1$.*

To prove this, consider firstly the form ω of order 1, say $\omega = a_i\,dx^i$. Then $d\omega = 0$ implies

$$\frac{\partial a_i}{\partial x^j} - \frac{\partial a_j}{\partial x^i} = 0.$$

This condition, evidently, is sufficient (and necessary) for the existence of such a function f that $a_i = f_{x_i}$, i.e. $\omega = df$.

For the case of a second-order form $\omega = a_{ij}\,dx^i \wedge dx^j$ the condition $d\omega = 0$ implies the system

$$\frac{\partial a_{ij}}{\partial x^k} + \frac{\partial a_{jk}}{\partial x^i} + \frac{\partial a_{ki}}{\partial x^j} = 0. \tag{7}$$

As ω_1 we take the first-order form $\omega_1 = b_i\,dx^i$. The equation $\omega = d\omega_1$ can be represented as

$$\frac{\partial b_j}{\partial x^i} - \frac{\partial b_i}{\partial x^j} = a_{ij}.$$

We rewrite this system in a definite order. First we write the equation containing the derivatives of b_1, then the remaining equations with derivatives of b_2 and so on. We get

$$\begin{aligned} \frac{\partial b_1}{\partial x^2} - \frac{\partial b_2}{\partial x^1} &= a_{21}, \\ &\cdots \\ \frac{\partial b_{n-1}}{\partial x^n} - \frac{\partial b_n}{\partial x^{n-1}} &= a_{nn-1}. \end{aligned} \tag{8}$$

We shall solve this system step by step from the bottom. We take an arbitrary C^2 function as b_n. From the last equation we can find b_{n-1} by integrating with respect to x^n. Substitute the functions b_n and b_{n-1} into the previous equation. Suppose that we already found the functions $b_n, \dots, b_i$. Let us find b_{i-1}. This function is included in a group of equations of (8), namely

$$\begin{aligned} \frac{\partial b_{i-1}}{\partial x^i} - \frac{\partial b_i}{\partial x^{i-1}} &= a_{ii-1}, \\ &\cdots \\ \frac{\partial b_{i-1}}{\partial x^n} - \frac{\partial b_n}{\partial x^{i-1}} &= a_{ni-1}. \end{aligned} \tag{9}$$

This system is integrable if the compatibility conditions are satisfied: for any $\alpha, \beta \geq i$ the difference of mixed derivatives of b_{i-1} is zero:

$$0 = \frac{\partial^2 b_{i-1}}{\partial x^\alpha\,\partial x^\beta} - \frac{\partial^2 b_{i-1}}{\partial x^\beta\,\partial x^\alpha} = \frac{\partial a_{\alpha i-1}}{\partial x^\beta} - \frac{\partial a_{\beta i-1}}{\partial x^\alpha} + \frac{\partial}{\partial x^{i-1}}\left(\frac{\partial b_\alpha}{\partial x^\beta} - \frac{\partial b_\beta}{\partial x^\alpha}\right). \tag{10}$$

Since $\alpha, \beta \geq i$, the b_α and b_β are already known and satisfy the equation

$$\frac{\partial b_\alpha}{\partial x^\beta} - \frac{\partial b_\beta}{\partial x^\alpha} = a_{\beta\alpha}.$$

Therefore, on the right of (10) we obtain

$$\frac{\partial a_{\alpha i-1}}{\partial x^\beta} + \frac{\partial a_{i-1\beta}}{\partial x^\alpha} + \frac{\partial a_{\beta\alpha}}{\partial x^{i-1}} = 0,$$

which is zero by virtue of system (7). Therefore, there is a function b_{i-1} satisfying (9). The sequence of functions $b_n, \ldots, b_i, \ldots, b_1$ provides the solution of (8). So, for the case of the second-order form ω the statement is proved.

The same method is applicable to the form

$$\omega = a_{i_1 \ldots i_{p+1}} dx^{i_1} \wedge \ldots \wedge dx^{i_{p+1}}.$$

The condition $d\omega = 0$ implies the system

$$\frac{\partial a_{i_1 \ldots i_{p+1}}}{\partial x^j} - \frac{\partial a_{ji_2 \ldots i_{p+1}}}{\partial x^{i_1}} - \cdots = 0. \tag{11}$$

Consider the form $\omega_1 = b_{i_1 \ldots i_p} dx^{i_1} \wedge \ldots \wedge dx^{i_p}$ of order p. The equation $\omega = d\omega_1$ is equivalent to the system

$$\frac{\partial b_{i_1 \ldots i_p}}{\partial x^j} - \frac{\partial b_{ji_2 \ldots i_p}}{\partial x^{i_1}} - \cdots = a_{ji_1 \ldots i_p}. \tag{12}$$

Introduce a multi-index $i = (i_1 \ \ldots \ i_p)$, where $i_1 \ \ldots \ i_p$. It is possible to compare multi-indices.

We say that $i < j$ if $i_1 < j_1$. In the case of $i_1 = j_1$ we set $i < j$ if $i_2 < j_2$ and so on. We put the coefficients $b_{i_1 \ldots i_p}$ into the sequence taking b_i earlier then b_j if $i < j$.

We also rewrite system (12) in a definite order. At first, consider the group of equations containing the derivative of b_1. These derivatives are with respect to $x^{p+1}, \ldots, x^n$. Then, select the group containing the derivatives of $b_{1 \ldots p-1p+1}$ from the remaining equations. They are the derivatives with respect to $x^{p+2}, \ldots, x^n$. The derivative of this function with respect to x^p is already included in the first group of equations. We continue this process for the next functions b_i. For every function $b_{i_1 \ldots i_p}$ the corresponding group of equations contains the derivatives of this function with respect to x^α only for $\alpha > i_p$. Indeed, if $\alpha < i_p$ then the equation

$$\frac{\partial b_{i_1 \ldots i_p}}{\partial x^\alpha} - \cdots - \frac{\partial b_{i_1 \ldots \alpha}}{\partial x^{i_p}} = a_{\alpha i_1 \ldots i_p}$$

is already included into the previous group corresponding to the function $b_{i_1 \ldots \alpha}$, since $(i_1 \ldots \alpha) < (i_1 \ldots i_p)$.

Therefore, we may assume that $\alpha > i_p$. This implies that the group of equations corresponding to $b_{i_1 \ldots i_p}$ includes the functions of greater multi-index.

The groups of equations contain, in general, a different number of equations. The function $b_{i_1 \ldots i_{p-1} n}$ has no corresponding group. Therefore, we can take the function

$b_{i_1\dots i_p\, n}$ to be arbitrary requiring only class C^2 regularity. Then we solve the system (12) ordered as above group by group starting from the bottom. Show that every group satisfies the compatibility condition. Suppose that we have already found the function b_j of the multi-index j, where $j > i$. Let us show that the function b_i exists, too. In the group which corresponds to $b_{i_1\dots i_p}$ we consider the equations

$$\frac{\partial b_{i_1\dots i_p}}{\partial x^\alpha} - \sum_{r=1}^{p} \frac{\partial b_{i_1\dots\alpha^r\dots i_p}}{\partial x^{i_r}} = a_{\alpha i_1\dots i_p},$$

$$\frac{\partial b_{i_1\dots i_p}}{\partial x^\beta} - \sum_{r=1}^{p} \frac{\partial b_{i_1\dots\beta^r\dots i_p}}{\partial x^{i_r}} = a_{\beta i_1\dots i_p};$$

where $\alpha, \beta > i_p$. The compatibility condition has the form

$$0 = \frac{\partial^2 b_i}{\partial x^\alpha \partial x^\beta} - \frac{\partial^2 b_i}{\partial x^\beta \partial x^\alpha}$$

$$= \frac{\partial a_{\alpha i_1\dots i_p}}{\partial x^\beta} - \frac{\partial a_{\beta i_1\dots i_p}}{\partial x^\alpha} + \sum_{r=1}^{p} \frac{\partial}{\partial x^{i_r}} \left(\frac{\partial b_{i_1\dots\alpha^r\dots i_p}}{\partial x^\beta} - \frac{\partial b_{i_1\dots\beta^r\dots i_p}}{\partial x^\alpha} \right). \tag{13}$$

where the r which are superscripts to α and β mean that the corresponding index takes the r-th position. In parenthesis we have the derivatives of functions of multi-indices $(i_1\dots\vee^r\dots i_p\alpha)$ and $(i_1\dots\vee^r\dots i_p\beta)$ which are greater than $(i_1\dots i_p)$. Therefore, by virtue of the inductive hypothesis the equations

$$\frac{\partial b_{i_1\dots\alpha^r\dots i_p}}{\partial x^\beta} - \frac{\partial b_{i_1\dots\beta^r\dots i_p}}{\partial x^\alpha} = \sum_{k=1,k\neq r}^{p} \frac{\partial b_{i_1\dots\alpha^k\dots\beta^r\dots i_p}}{\partial x^{i_k}}$$

are satisfied.

On the right of (13) we obtain

$$\frac{\partial a_{\alpha i_1\dots i_p}}{\partial x^\beta} - \frac{\partial a_{\beta i_1\dots i_p}}{\partial x^\beta} - \frac{\partial a_{\beta i_1\dots\alpha^r\dots i_p}}{\partial x^{i_r}} + \sum_{i_r, i_k} \frac{\partial^2 b_{i_1\dots\alpha^k\,\beta^r\dots i_p}}{\partial x^{i_r}\partial x^{i_k}}.$$

The last sum in the equation above is zero due to the skew-symmetry of $b_{i_1\dots i_p}$ with respect to indices. The sum of the remaining terms is zero because of (11). Therefore, the group of equations on b_i is compatible. If we consequently find all the b_i then we find the form ω_1. This completes the proof of the Poincaré lemma.

Observe that ω_1 is not determined uniquely. For instance, it is always possible to add the form $d\omega_2$.

The form ω having a zero exterior differential is called an *exact form*.

2.10 The Exterior Codifferential

In a Riemannian space the operator δ of exterior codifferentiation is defined which puts the exterior form ω of order p and the exterior form $\delta\omega$ of order $p-1$ into correspondence. Let ∇ denote the covariant derivative in Riemannian space and $\nabla_k a_{i_1 \dots i_p}$ be the covariant derivative of the tensor $a_{i_1 \dots i_p}$. Set

$$\nabla^i a_{i_1 \dots i_p} = g^{ik} \nabla_k a_{i_1 \dots i_p}.$$

Since $\nabla^i a_{ii_1 \dots i_{p-1}}$ is a skew-symmetric tensor of order $p-1$, we are able to define the following exterior form:

$$\delta\omega = -\sum_{i_1 \dots i_{p-1}} \frac{1}{p!} \nabla^i a_{ii_1 \dots i_{p-1}} dx^i_1 \wedge \dots \wedge dx^{i_{p-1}}.$$

The codifferential satisfies the Poincaré lemma $\delta^2\omega = 0$.

The proof of the latter statement is more complicated than for the case of the differential d. It is based on the properties of the Riemannian tensor. We produce it, first, in the case of the second-order form $\omega = a_{ij}\, dx \wedge dx$. Then $\delta^2\omega$ is a scalar function. We have

$$\delta^2\omega = g^{jl} \nabla_l (g^{ik} \nabla_k a_{ij}) = g^{jl} g^{ik} \nabla_l \nabla_k a_{ij} = -g^{jl} g^{ik} \nabla_l \nabla_k a_{ji}. \tag{1}$$

Make the index change $l \to k \to l$ and then change the order of covariant differentiation. If we take into account that the difference of second-order covariant derivatives of a tensor has an expression in terms of the tensor contractions with a Riemannian tensor then we see that

$$\delta^2\omega = -g^{jk} g^{il} \nabla_k \nabla_l a_{ji} = -g^{jk} g^{il} (\nabla_l \nabla_k a_{ji} + R^{\alpha}_{\cdot jkl} a_{\alpha i} + R^{\alpha}_{\cdot ikl} a_{j\alpha}). \tag{2}$$

Consider the following expression (we denote it by A_1):

$$A_1 = R^{\alpha}_{\cdot jkl} a_{\alpha i} g^{ik} g^{jl} = R_{\beta jkl} a_{\alpha i} g^{\alpha\beta} g^{jk} g^{il}.$$

Interchanging the roles of α and i, l and β, j and k we obtain

$$A_1 = R_{lkj\beta} a_{i\alpha} g^{il} g^{kj} g^{\alpha\beta} = -R_{j\beta lk} a_{\alpha i} g^{\alpha\beta} g^{il} g^{kj} = -A_1.$$

Therefore $A_1 = 0$. In an analogous way we find that the second contraction with a Riemannian tensor is zero:

$$\begin{aligned} A_2 = g^{jk} g^{il} R^{\alpha}_{\cdot ikl} a_{j\alpha} &= R_{\beta ikl} a_{j\alpha} g^{\beta\alpha} g^{jk} g^{il} = R_{\beta jkl} a_{i\alpha} g^{\beta\alpha} g^{ik} g^{jl} \\ &= R_{\beta jlk} a_{i\alpha} g^{\beta\alpha} g^{il} g^{jk} = R_{\beta jkl} a_{\alpha i} g^{\beta\alpha} g^{il} g^{jk} - A_1 = 0. \end{aligned}$$

So, we have

$$\delta^2\omega = -g^{jk} g^{il} \nabla_l \nabla_k a_{ji} = -g^{ik} g^{jl} \nabla_l \nabla_k a_{ij},$$

which differs from the initial expression only by the sign. Therefore, $\delta^2\omega = 0$.

Now, consider the form $\omega = a_{iji_1 \dots i_p} dx^i \wedge \dots dx^{i_p}$ of order $p+2$, where $p \geq 1$. Let $b_{i_1 \dots i_p}$ be the coefficients of $\delta^2\omega$. We have

$$\begin{aligned} b_{i_1 \dots i_p} &= g^{ik} g^{jl} \nabla_l \nabla_k a_{iji_1 \dots i_p} = -g^{ik} g^{jl} \nabla_l \nabla_k a_{jii_1 \dots i_p} \\ &= -g^{jk} g^{il} \nabla_k \nabla_l a_{jii_1 \dots i_p} = -g^{jk} g^{il} (\nabla_l \nabla_k a_{jii_1 \dots i_p} \\ &\quad + R^{\alpha}_{\cdot jkl} a_{\alpha i i_1 \dots i_p} + R^{\alpha}_{\cdot ikl} a_{j\alpha i_1 \dots i_p} + R^{\alpha}_{\cdot i_1 kl} a_{ji\alpha \dots i_p} + \dots + R^{\alpha}_{\cdot i_p kl} a_{ji \dots \alpha}). \end{aligned} \tag{3}$$

Just in the same way as for the second-order form we can see that the contraction of $g^{jk}g^{il}$ with the first two terms on the right of (3), where the Riemannian tensor involved, is zero. Consider the following expression:

$$A_3 = g^{jk}g^{il}R^{\alpha}_{\cdot i_1kl}a_{ji\alpha\ldots i_p} = R_{\beta i_1kl}a_{ji\ldots i_p}g^{\beta\alpha}g^{jk}g^{il}$$
$$= -R_{i_1\beta kl}a_{ji\alpha i_2\ldots i_p}g^{\beta\alpha}g^{jk}g^{il}.$$

Let us make the index change as $\beta \to k \to l \to \beta$ and $\alpha \to j \to i \to \alpha$. Since $a_{i\alpha ji_2\ldots i_p} = a_{ij\alpha i_2\ldots i_p}$,

$$A_3 = -R_{i_1kl\beta}a_{ji\alpha\ldots i_p}g^{\beta\alpha}g^{jk}g^{il}.$$

Making the same index change and adding the obtained expressions, we find

$$3A_3 = -(R_{i_1\beta kl} + R_{i_1kl\beta} + R_{i_1l\beta k})a_{ji\alpha i_2\ldots i_p}g^{\beta\alpha}g^{jk}g^{il}.$$

By virtue of the Bianchi identity we see that $A_3 = 0$. In an analogous way we obtain zero in all consequent terms of (3). Therefore,

$$b_{i_1\ldots i_p} = -g^{jk}g^{il}\nabla_l\nabla_k a_{jii_1\ldots i_p} = -g^{jk}g^{il}\nabla_l\nabla_k a_{ij\ldots i_p}.$$

Comparing this and the initial expression for $b_{i_1\ldots i_p}$, we see that the coefficients $b_{i_1\ldots i_p}$ are zero. So, $\delta^2\omega = 0$. One can get a simpler proof of this fact by means of the Hodge star operator.

2.11 Some Formulas for the Exterior Differential

It is easy to represent the integrability condition of the Pfaff equation in terms of the exterior differential and exterior product. The following theorem holds.

Cartan Lemma In order for the second-order exterior form $F = a_{ij}\,dx^i \wedge dx^j$ to be equal to zero by virtue of the equation $f = a_i\,dx^i = 0$ it is necessary and sufficient that there is a linear form φ such that

$$F = f \wedge \varphi.$$

To prove this result, we go from the basis dx^i to the basis which includes the linear form f. Suppose that $f_1 \ldots f_n$ is the required basis, where $f_1 = f$.

Consider

$$F = b_{ij}f_i \wedge f_j = \sum_{j=1}^{n} b_{1j}f_1 \wedge f_j + \sum_{\mu,\nu=2}^{n} b_{\mu\nu}f_\mu \wedge f_\nu. \qquad (1)$$

Since $F = 0$ for $f_1 = 0$ and since $f_\mu \wedge f_\nu$ are linearly independent second-order forms, $b_{\mu\nu} = 0$. Therefore, $F = f_1 \wedge \sum_{j=1}^{n} b_{1j}f_j$. Introduce the linear form

$$\varphi = \sum_{j=1}^{n} b_{1j}f_j.$$

Then (1) can be written as $F = f \wedge \varphi$.

Let us be given the Pfaff equation $\omega = 0$. We can represent the bilinear Frobenius form $d\omega(\delta) - \delta\omega(d)$ as the exterior differential of ω if we set

$$dx^i \delta x^j - \delta x^i dx^j = dx^i \wedge dx^j,$$

i.e.

$$d\omega(\delta) - \delta\omega(d) = d\omega.$$

The Pfaff equation $\omega = 0$ is integrable if and only if the bilinear Frobenius form, i.e. $d\omega$, is zero by virtue of the equation $\omega = 0$. The lemma just proved implies that there is a linear form φ such that

$$d\omega = \omega \wedge \varphi. \tag{2}$$

This is the necessary and sufficient condition for the Pfaff equation $\omega = 0$ to be integrable. In other words, *in the integrable case the form $d\omega$ is divisible by ω*. This condition can also be represented as $\omega \wedge d\omega = 0$.

For every linear Pfaff form $\omega = a_i dx^i$ we may adjoin the vector field $\tau = \{a_i\}$. Let $\mathbf{N} = \{N_i\}$ be a unit vector field of principal normals of the streamlines of the field τ and k be the curvature of these streamlines. We may also adjoin the linear form $\eta = N_i dx^i$ for this field. Let τ be the unit vector field. Let us show that *if τ is holonomic* or, equivalently, the equation $\omega = 0$ is totally integrable *then it is possible to set $\varphi = k\eta$* in equation (2), i.e.

$$d\omega = k\omega \wedge \eta.$$

This equation can be regarded as the formula analogous to the Hamilton formula from Section 1.3. The form $k\eta$ is

$$k\eta = \frac{\partial a_i}{\partial x^j} a_j dx^i.$$

Find the exterior product

$$\omega \wedge k\eta = a_k dx^k \wedge \frac{\partial a_i}{\partial x^j} a_j dx^i = \frac{1}{2}\left(a_k \frac{\partial a_i}{\partial x^j} - a_i \frac{\partial a_k}{\partial x^j}\right) a_j dx^k \wedge dx^i.$$

Prove the equality of $d\omega$ and $\omega \wedge k\eta$ in a special choice of coordinate system. Set the coordinate curve x^n to be tangent to τ at some fixed point P_0 while the other coordinate curves are orthogonal to τ. Then at P_0 we have $a_1 = \cdots = a_{n-1} = 0$, $a_n = 1$, $\partial a_n/\partial x^i = 0$. The form ω at P_0 is

$$d\omega = \frac{1}{2}\sum_{\alpha,\beta<n}\left(\frac{\partial a_\beta}{\partial x^\alpha} - \frac{\partial a_\alpha}{\partial x\beta}\right) dx^\alpha \wedge dx^\beta + \sum_{\alpha=1}^{n-1}\left(\frac{\partial a_n}{\partial x^\alpha} - \frac{\partial a_\alpha}{\partial x^n}\right) dx^\alpha \wedge dx^n.$$

Since the conditions $\frac{\partial a_\alpha}{\partial x^\beta} - \frac{\partial a_\beta}{\partial x^\alpha} = 0$ $\alpha, \beta \leq n-1$ holds (see (14) of Section 2.7) and $\frac{\partial a_n}{\partial x^n} = 0$,

$$d\omega = -\sum_{\alpha=1}^{n-1} \frac{\partial a_\alpha}{\partial x^n}\, dx^\alpha \wedge dx^n. \tag{3}$$

Next we consider $\omega \wedge k\eta$. Since at P_0 $a_k = a_i = 0$ for $i, k < n$,

$$\omega \wedge k\eta = \sum_{i=1}^{n-1} \frac{\partial a_i}{\partial x^n}\, dx^n \wedge dx^i. \tag{4}$$

Comparing (3) and (4) we conclude that $\omega \wedge k\eta$ and $d\omega$ coincide with each other at P_0. The coincidence of the forms does not depend on the choice of coordinate system because of their invariant meaning. Therefore, the equality $d\omega = k\omega \wedge \eta$ is proved.

For every k-th order form $\omega = a_{i_1 \dots i_k} dx_1^i \wedge \dots \wedge dx^{i_k}$ it is possible to put into correspondence its value on k vectors $\mathbf{Y}_1, \dots, \mathbf{Y}_k$, namely, the contraction of the tensor $a_{i_1 \dots i_k}$ with the components of a multi-vector generated by $\mathbf{Y}_1, \dots, \mathbf{Y}_k$. Denote this value by $\omega(\mathbf{Y}_1, \dots, \mathbf{Y}_k)$. Then

$$\omega(\mathbf{Y}_1, \dots, \mathbf{Y}_k) = a_{i_1 \dots i_k} \begin{vmatrix} Y_1^{i_1} & \dots & Y_1^{i_k} \\ \dots & \dots & \dots \\ Y_k^{i_1} & \dots & Y_k^{i_k} \end{vmatrix}.$$

Let us state a useful formula. Let $\omega = a_i dx^i$ be a linear form, τ the corresponding unit vector field, τ_α and τ_β the fields orthogonal to τ. Then

$$d\omega(\tau_\alpha, \tau_\beta) = (\tau, \nabla_{\tau_\alpha} \tau_\beta - \nabla_{\tau_\beta} \tau_\alpha). \tag{5}$$

Indeed we have

$$\begin{aligned} d\omega(\tau_\alpha, \tau_\beta) &= \frac{1}{2}\left(\frac{\partial a_i}{\partial x^j} - \frac{\partial a_j}{\partial x^i}\right) \begin{vmatrix} \tau_\alpha^i & \tau_\alpha^j \\ \tau_\beta^i & \tau_\beta^j \end{vmatrix} \\ &= \frac{1}{2}\left\{-a_i \frac{\partial \tau_\alpha^i}{\partial x^j} \tau_\beta^j + a_j \frac{\partial \tau_\beta^j}{\partial x^i} \tau_\alpha^i + a_i \frac{\partial \tau_\beta^i}{\partial x^j} \tau_\alpha^j - a_j \frac{\partial \tau_\alpha^j}{\partial x^i} \tau_\beta^i\right\} \\ &= (\tau, \nabla_{\tau_\alpha} \tau_\beta - \nabla_{\tau_\beta} \tau_\alpha), \end{aligned}$$

where τ_α^i stands for the components of τ_α. The formula (5) is established.

2.12 Simplex, the Simplex Orientation and the Induced Orientation of a Simplex Boundary

Let us be given a set of $(n+1)$ points in Euclidean space E^n: $M_0, M_1, \dots, M_n$. Let $\mathbf{r}_i$ be the position vectors. The set of end-points of all vectors

$$\mathbf{r} = x^0 r_0 + x^1 r_1 + \dots + x^n r_n,$$

where the numbers x^i satisfy the condition

$$x^0 + x^1 + \cdots + x^n = 1, \quad 0 \le x^i \le 1,$$

is called the *n-simplex*. We shall denote the simplex by t^n pointing out the order of vertices if necessary: $t^n(M_0, M_1, \ldots, M_n)$. Define the simplex orientation as the orientation in E^n generated by the basis $M_0M_1, \ldots, M_0M_n$. The odd permutation of vertices make the orientation opposite, while the even one preserves the orientation. We shall denote the simplex of orientation opposite to t^n by $-t^n$.

Select in a given simplex any p vertices and form a simplex $t^p(M_{i_0}, M_{i_1}, \ldots, M_{i_p})$. This subsimplex is called *the face*.

Consider the $(n-1)$-faces. Denote by t_i^{n-1} or $T_i^{n-1}(M_0, \ldots, \vee^i, \ldots M_n)$ the face obtained by deleting M_i, where $\vee^i$ means the omission of M_i.

The orientation of a simplex $(M_0, \ldots, \vee^i, \ldots M_n)$ is called *induced by the simplex* $t^n(M_0, \ldots, M_n)$ if the simplex $(M_i, M_0, \ldots, M_n)$ has the same orientation as t^n. The simplex t^{n-1} of orientation induced by t^n has the form $t^{n-1} = (-1)^i(M_0, \ldots, \vee^i, \ldots M_n)$.

2.13 The Simplicial Complex, the Incidence Coefficients

First, we consider the simplicial complex in Euclidean space E^m. *The simplicial complex in E^m of order n* is a finite or countable set of simplices satisfying the following conditions

(1) Together with the simplex t^p, this set contains all its faces including the vertices.
(2) Every p-simplex t^p of a simplicial complex is a face of some n-simplex t^n (*the homogenity property*).
(3) The intersection of any two simplices $t^p \bigcap t^q$ either is empty or is a simplex of a given complex.
(4) Every two simplices can be joined with the chain of simplices having link-wise non-empty intersections (*the connectivity property*).
(5) Any bounded domain in space intersects a finite number of simplices.

The set of points in E^m which belongs to the complex is called *the polyhedron*. We say that the smooth manifold M^n is triangulated if it is mapped onto some polyhedron homeomorphically. We shall call the image of a simplex in M^n *the curvilinear simplex*. The fundamental hypothesis of combinatorial topology ("Hauptvermutung") asserts that any two triangulations of homeomorphic polyhedra have isomorphic partitions. In recent years this hypothesis has been refuted.

Suppose that every simplex in a simplicial complex is oriented. Introduce the incidence coefficient of the two simplices t^n and t^{n-1}. We say that the incidence coefficient (t^nt^{n-1}) is zero if t^{n-1} is not a face of t^n. If t^{n-1} is a face of t^n then $(t^nt^{n-1}) = 1$ for the case when the orientation in t^{n-1} induced by t^n coincides with the given one and $(t^nt^{n-1}) = -1$ for the opposite case. Let t^{n-2} be some subsimplex in t^n

and t_0^{n-1}, t_1^{n-1} be the $(n-1)$-simplices in t^n which include t^{n-1}. Then for any choice of orientation of simplices the following holds:

$$(t^n t_0^{n-1})(t_0^{n-1} t^{n-2}) + (t^n t_1^{n-1})(t_1^{n-1} t^{n-2}) = 0. \tag{1}$$

Set, for example, that t^{n-2} is $(M_2 \dots M_n)$. The value of each term of the above will not change if we change the orientations in t_0^{n-1} and t_1^{n-1}. Suppose that t_0^{n-1} does not contain the vertex M_0, while t_1^{n-1} does not contain the vertex M_1. Orient t_0^{n-1} and t_1^{n-1} in such a way that

$$(t_0^{n-1} t^{n-2}) = 1, \quad (t_1^{n-1} t^{n-2}) = 1.$$

Then the simplices t_0^{n-1} and t_1^{n-1} have the forms

$$t_0^{n-1} = (M_1 M_2 \dots M_n),$$
$$t_1^{n-1} = (M_0 M_2 \dots M_n).$$

Under the conditions posed on the incidence coefficients the formula we need to prove has the form

$$(t^n t_0^{n-1}) + (t^n t_1^{n-1}) = 0.$$

This relation stays the same if we change the orientation of t^n into the opposite one. Therefore, we choose the orientation by setting, for instance,

$$t^n = (M_0 M_1 M_2 \dots M_n).$$

Then

$$(t^n t_0^{n-1}) = 1, \quad (t^n t_1^{n-1}) = -1.$$

Hence, formula (1) is true.

Consider now the simplicial complex. Let t^n and t^{n-2} be some selected simplices. Then

$$\sum_i (t^n t_i^{n-1})(t_i^{n-1} t^{n-2}) = 0, \tag{2}$$

where the summation is assumed over all $(n-1)$–simplices.

If t^{n-2} is not a subsimplex of t^n then every simplex t_i^{n-1} which includes t^{n-2} is not a face of t^n. In this case $(t^n t_i^{n-1}) = 0$ and (2) is true. Therefore, suppose that $t^{n-2} \subset t^n$. This simplex contains precisely two simplices t_i^{n-1} which include t^{n-2}. Hence, (2) is reduced to the formula (1) which has been proved already.

2.14 The Integration of Exterior Forms

Suppose that the n-dimensional manifold R^n endowed with curvilinear coordinates x^i and the form of order n is given:

$$\omega = a_{1\dots n}\, dx^1 \wedge \dots \wedge dx^n$$

or

$$\omega = \frac{1}{n!}\sum_{i_1\dots i_n} a_{i_1\dots i_n}\, dx^{i_1}\wedge\dots\wedge dx^{i_n}.$$

In a coordinate change the coefficients of the n-th order form gives the multiplier which is the Jacobian of the coordinate transformation $I\left(\frac{x^1\dots x^n}{u^1\dots u^n}\right)$. It is possible to define the integral of the form ω over $D\subset R^n$ in the usual way as the integral of $a_{1\dots n}$

$$\int_{R^n}\omega = \int_D a_{1\dots n}\, dx^1\dots dx^n.$$

The value of the form integral does not depend on the choice of the coordinate system. Indeed, with respect to a u^i coordinate system the form ω is

$$\omega = a_{1\dots n} I\left(\frac{x^i}{u^i}\right) du^1\wedge\dots\wedge du^n.$$

Let G be a domain in a space of $u^1,\dots,u^n$ which corresponds to D. Since

$$\int_D a_{1\dots n}\, dx^1\dots dx^n = \int_G a_{1\dots n} I\left(\frac{x^i}{u^i}\right) du^1\dots du^n,$$

the integral of the form depends on the form and the domain only.

Suppose now that the form ω of order p is given in a domain D. Let F^m be an m-dimensional surface represented as

$$x^i = f^i(y^1,\dots,y^m),\quad i=1,\dots,n.$$

On the surface F^m the differentials dx^i are some linear forms depending on dy^j, $j=1,\dots,m$. Therefore, on the surface F^n some new exterior form of order m is induced. We shall call this form *the induced one* and denote it by $\bar\omega$:

$$\bar\omega = \frac{1}{p!}\sum_{\substack{i_1\dots i_p\\ j_1\dots j_m}} a_{i_1\dots i_p}\frac{\partial f^{i_1}}{\partial y^{j_1}}\cdots\frac{\partial f^{i_p}}{\partial y^{j_m}} dy^{j_1}\wedge\dots\wedge dy^{j_m}.$$

Thus, form coefficients are, evidently, skew-symmetric with respect to indices $j_1,\dots,j_m$. The integral of ω over the m-dimensional surface F^m is, by definition, the integral of $\bar\omega$ over F^m

$$\int_{F^m}\omega = \int_{F^m}\bar\omega.$$

Consider, as a particular case, the form ω of order $n-1$ in some n-dimensional oriented domain D. Let Γ be a smooth boundary of D endowed with the orientation induced from D. In D the form $d\omega$ of order n is also defined. Hence, both of the following integrals are defined:

$$\int_{\Gamma} \omega \quad \text{and} \quad \int_{D} d\omega.$$

Denote the boundary of D by ∂D.

The generalized Stokes' formula holds:

$$\int_{\partial D} \omega = \int_{D} d\omega.$$

This formula establishes the close relation between the notions of exterior differential and boundary.

We prove this formula in the case when R^n is the affine space of coordinates $x^1, \ldots, x^n$ and D is an n-simplex defined by the inequalities

$$x^1 + x^2 + \cdots + x^n \leq 1, \quad 0 \leq x^i \leq 1.$$

Let $A_0, A_1, \ldots, A_n$ be the simplex vertices. Their ordering gives the simplex orientation. We shall represent the simplex as $(A_0, \ldots, A_n)$. We are able to represent any subsimplex in an analogous manner. The simplex boundary is the set of all its signed $(n-1)$-subsimplices, namely

$$\partial D = \sum_{i=1}^{n} (-1)^i (A_0, \ldots, A_{i-1}, A_{i+1}, \ldots, A_n).$$

At any point of a boundary simplex one of its coordinates satisfies the equation $x^i = 0$. Let ω be the form of order $(n-1)$ in D:

$$\begin{aligned}\omega &= \sum_{i_1 \ldots i_{n-1}} a_{i_1 \ldots i_{n-1}}\, dx^{i_1} \wedge \ldots \wedge dx^{i_{n-1}} \\ &= a_{2 \ldots n-1}\, dx^2 \wedge \ldots \wedge dx^n + \cdots + a_{12 \ldots n-1}\, dx^1 \wedge \ldots \wedge dx^{n-1}.\end{aligned}$$

Each term is itself some exterior form of order $n-1$. Let us prove Stokes' formula for each of them. Take, for instance, the form

$$\omega_1 = a_{2 \ldots n-1}\, dx^2 \wedge \ldots \wedge dx^n.$$

On the simplex face D_i, which is given by $x^i = 0$, the differential $dx^i = 0$. Therefore, on the simplex D_i the form ω_1 is zero if $i \geq 2$. Hence, ω_1 is different from zero on D_1 and D_0 only when

$$\int_{\partial D} \omega_1 = \int_{D_1} \omega_1 + \int_{D_0} \omega_1.$$

On the face $D_0 = (A_1, \ldots, A_n)$ we have $x^1 = 1 - x^2 - \cdots - x^n$. Settle the correspondence between the faces D_0 and D_1 by the equality of coordinates, i.e. by means

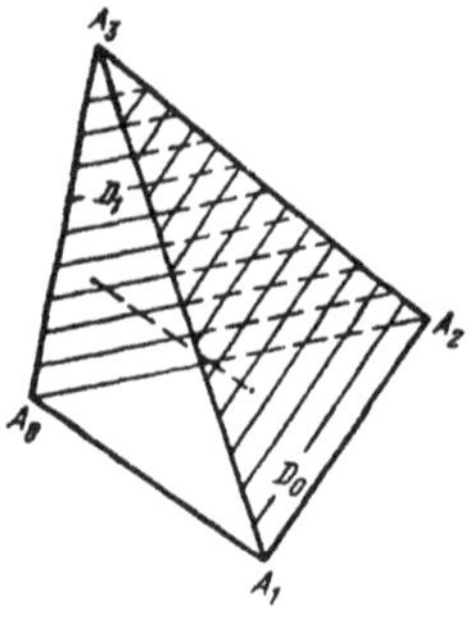

FIGURE 24

of projection (see Fig. 24). The simplex D_1 is given as $D_1 = (-1)(A_0A_2 \ldots A_n)$. Therefore, the orientation of D_1 defined by coordinates $x^2, \ldots, x^n$ is opposite to its orientation as part of the boundary of D. So, we have

$$\int_{\partial D} \omega_1 = -\int_{D_0} a_{2\ldots n}(0, x^2, \ldots, x^n)\, dx^2 \wedge \ldots \wedge dx^n$$
$$+ \int_{D_0} a_{2\ldots n}(1 - x^2 - \ldots, x^2, \ldots, x^n)\, dx^2 \wedge \ldots \wedge dx^n$$
$$= \int_{D_0} \int_0^{x_0^1} \frac{\partial}{\partial x^1} a_{2\ldots n}(x^1, x^2, \ldots, x^n)\, dx^1 \wedge dx^2 \wedge \ldots \wedge dx^n,$$

where $x_0^1 = 1 - x^2 - \cdots - x^n$. The latter expression is the integral over D. On the other hand, consider the exterior differential of ω_1:

$$d\omega_1 = da_{2\ldots n-1} \wedge dx^2 \wedge \ldots \wedge dx^n = \frac{\partial}{\partial x^1} a_{2\ldots n-1}\, dx^1 \wedge dx^2 \wedge \ldots \wedge dx^n.$$

Hence

$$\int_{\partial D} \omega_1 = \int_D d\omega_1.$$

In an analogous manner one can prove this formula for any other ω_i. For the case of a curvilinear simplex in some manifold R^n the proof can be reduced to the case of a simplex in affine n-dimensional space.

Further on we assume that D is triangulated, and also that the boundary of D is represented as $(n-1)$-simplices of triangulation. The theorem that any smooth manifold can be triangulated is known. The integral of $d\omega$ over each simplex of triangulation is equal to the integral of ω over the $(n-1)$-simplices of the boundary. Let us consider the sum of those integrals. If the n-simplices are oriented consistently then for any $(n-1)$-simplex in a common boundary of two distinct n-simplices the orientations induced on it from those two simplices are opposite. Therefore, in the sum under consideration the integrals over $(n-1)$-simplices interior to D cancel each other out. The remaining integrals are ones over the $(n-1)$-simplices on the boundary of D. This completes the proof of Stokes' formula.

2.15 Homology and Cohomology Groups

Let us be given a simplicial partition of a topological space. Each simplex of the partition is already endowed with an orientation. Denote the i-th simplex of dimension r by t_i^r. Let us form a chain of simplices as

$$x^r = \sum_i a_i t_i^r,$$

where a_i are some integers. *A boundary of the chain r* is a sum of the boundaries of simplices t_i^r if we take them with coefficients a_i:

$$dx^r = \sum_i a_i \, dt_i^r.$$

To find the boundary of a single simplex we use the incidence coefficients. Then the boundary of t_i^r is the following $(r-1)$-dimensional chain:

$$dt_i^r = \sum_j (t_i^r t_j^{r-1}) t_j^{r-1}.$$

The boundary taken twice is zero, i.e. the boundary of a boundary is zero. It is sufficient to prove this for the boundary of a single simplex. We have

$$d\,dt_i^r = \sum_j (t_i^r t_j^{r-1}) \sum_k (t_j^{r-1} t_k^{r-2}) t_k^{r-2} = \sum_k \sum_j (t_i^r t_j^{r-1})(t_j^{r-1} t_k^{r-2}) t_k^{r-2}.$$

In Section 2.13 we proved the equality

$$\sum_j (t_i^r t_j^{r-1})(t_j^{r-1} t_k^{r-2}) = 0.$$

The chain x^r with zero boundary is called *closed* or *the cycle*, i.e. for the case of a closed chain $dx^r = 0$. The chain x^r of dimension r which is the boundary of an $(r+1)$-dimensional chain y^{r+1} is called *homologous to zero*: $x^r = dy^{r+1}$.

The sum of two chains of the same dimension

$$x^r = \sum_i a_i t_i^r, \quad y^r = \sum_i b_i t_i^r$$

is the chain

$$x^r + y^r = \sum_i (a_i + b_i) t_i^r.$$

The sum of two cycles is a cycle. Therefore, the set of all cycles of dimension r form the Abelian group Z_r with respect to the adding operation as above. The quotient group of Z_r with respect to the subgroup of cycles homologous to zero is called *the r-dimensional Betti group*. If T_r is a subgroup of Z_r formed by the cycles homological to zero, then the Betti group H_r is

$$H_r = Z_r / T_r.$$

In an analogous way the cohomology group can be defined. For each cell t_i^r the coboundary is the $(r+1)$-dimensional chain of the form

$$\delta t_i^r = \sum_i (t_i^r t_i^{r+1}) t_i^{r+1}.$$

The chain coboundary is the additive sum of coboundaries of chain cells

$$\delta x^r = \sum_i a_i \delta t_i^r.$$

The chain x^r which is a coboundary of some $(r-1)$-dimensional chain y^{r-1} is called *cohomologous to zero*. If a coboundary of the chain x^r is zero then this chain is called a cocycle: $\delta x^r = 0$. The set of all cocycles, being endowed with the natural operation of adding, forms the additive group denoted by Z^r.

Let T^r be a subgroup of Z^r formed by the cocycles homologous to zero. Then the quotient group of Z^r with respect to T^r is called *the cohomology group*:

$$H^r = Z^r / T^r$$

Consider, for instance, the cohomology group $H_1(S^1)$ with integer coefficients of the circle S^1. Divide S^1 with the points A, B, C into three arcs, namely $AB = y_1$, $BC = y_2$, $CA = y_3$. We assume that the given sequence of end-points of arcs y_i produces the orientation of the arc. The coboundary of A is formed by two arcs, namely y_1 and y_3 giving

$$\delta A = y_3 - y_1.$$

Analogously, $\delta B = y_1 - y_2$, $\delta C = y_2 - y_3$. Hence, the arcs y_3 and y_2 are cohomologous to y_1:

$$y_3 = y_1 + \delta A, \quad y_2 = y_1 + \delta A + \delta C.$$

Therefore, the group $H^1(S^1)$ is isomorphic to the group of integers Z.

There is a close relation between the space of different forms defined over the whole manifold and the cohomology groups of the manifold, namely the following *de Rham theorem* holds:

The quotient space of closed forms of order p with respect to the subspace of exact forms of order p is isomorphic to the p-dimensional cohomology group with real coefficients $H^p(M, R)$ of the manifold M.

2.16 Foliations on the Manifolds and the Reeb's Example

Let us define the notion of a foliation on the manifold. Let M^n be a regular Riemannian manifold of class C^r without a boundary. Suppose that M^n is decomposed into connected subsets which we shall call *the fibers of foliation* or *the leaves* and denote them by L. Let M^n be covered by the set of neighborhoods U_α each endowed with a coordinate system $x^1, \ldots, x^n$ (i.e. there is the C^r regular homeomorphism $\varphi_\alpha : U_\alpha \to V_\alpha \subset E^n$ of U_α onto some domain V_α in E^n) such that the intersection of each leaf L with U_α can be determined by the system of equations $x^{p+1} = \text{const}, \ldots, x^n = \text{const}$. In this case we say that on the manifold M^n *the C^r-differentiable foliation with the leaves of dimension p* is given. The number $n - p$ is called *the codimension* of the foliation. Each fiber is a submanifold of M^n. Thus, *the foliation is structured locally like a family of parallel planes of dimension p in Euclidean space E^n*. If M^n is an analytic manifold and φ_α are analytic functions then we say that the foliation is analytic.

Among the closed surfaces M^2 only the torus and the Klein bottle admit foliations without singular points. For $n = 3$ the foliations of class C^r exist on any manifold. However, the manifold containing an analytic foliation must satisfy the condition with respect to its fundamental group. Haefliger stated that *a compact manifold with a finite fundamental group does not admit analytic foliations of codimension 1.*

The analytic requirement in the theorem is essential. *Reeb constructed the foliation of class C^∞ on the three-dimensional sphere S^3*. We describe this foliation here after Lawson. At first, construct the foliation of the toroidal body K^3. The boundary of this body is a two-dimensional torus which we denote by T^2. It will be one of the fibers, moreover the limit fiber for the other fibers. As Lawson said figuratively, each fiber within K^3 is like a snake that eats its tail. While the phrase above expresses everything you need to understand, we continue the explanation. In the xy-plane we take a zone between two parallel straight lines $x = -1$ and $x = 1$. Construct the line foliation in it. Let us be given a curve $y = f(x)$ defined in $-1 < x < 1$. The function $f(x)$ is supposed to be of class C^∞, infinitely large when $|x| \to 1$ and for all derivatives $f^{(k)}(x)$

$$\lim_{|x| \to 1} f^{(k)}(x) = \infty, \quad k = 1, \ldots, \infty.$$

We define each line of zone foliation with the equation $y(x) = f(x) + c$, where c is an arbitrary real number (see Fig. 25). Revolving this foliation around the y-axis, we obtain the foliation of a cylindrical body. Let us map the cylindrical body onto the toroidal body K^3 in the following standard way. Consider the axial curve (the circle S^1) in the usual torus. We assume that circle radius is 1. Let s be the arc length of S^1 with respect to some fixed point P_0 as the initial one. Make the section of the

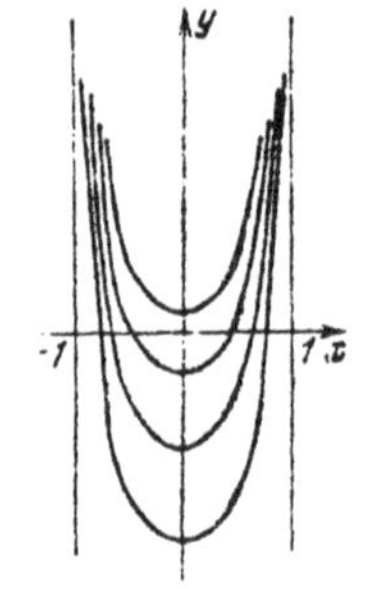

FIGURE 25

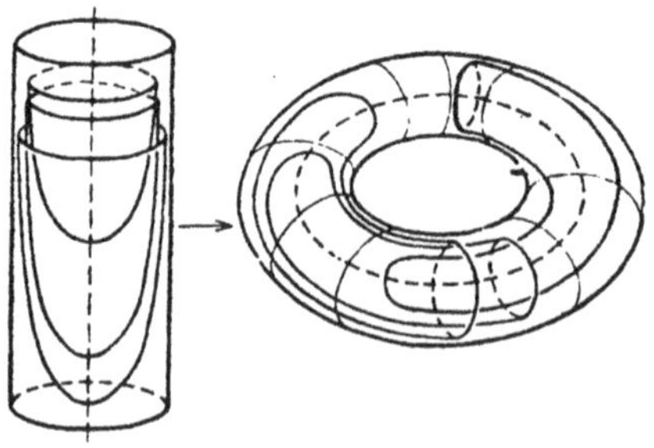

FIGURE 26

toroidal body K^3 through the point P in S^1 with a plane orthogonal to S^1 at P. Denote a disk obtained with that section by D^2. We can represent the toroidal body K^3 as a product $S^1 \times D^2$. Introduce the polar coordinates (ρ, θ) in D^2 taking P as a pole. In a section $y = \text{const}$ we have a disk. Introduce the polar coordinates (r, φ) in this disk with a pole at a point in the y-axis. Now define a mapping ψ of the cylindrical body onto K^3 setting $s = y$, $\rho = r$, $\theta = \varphi$. Every torus point will be covered infinitely many times under this mapping; also, if the points a_1 and a_2 of a cylindrical body go into the same point of K^3 then the difference of corresponding ordinates satisfies $y_1 - y_2 = 2\pi k$, where k is an integer, while the corresponding parameters (r, φ) coincide. Let us show that under this mapping the foliation of the cylindrical body goes into the of K^3 (see Fig. 26). Locally, the mapping ψ is a homeomorphism. Suppose that the fibers l_1 and l_2 of the cylindrical body went into the surfaces $\psi(l_1)$ and $\psi(l_2)$ which have a common point b. Then the inverse images of this point, say a_1 and a_2, differ in ordinates as $y_1 - y_2 = 2\pi k$, while the polar parameters are the same, $r_1 = r_2$, $\varphi_1 = \varphi_2$. The fibers l_1 and l_2 have been obtained in revolving the curves $\bar{y} = y(r) + c_1$ and $\tilde{y} = y(r) + c_2$ respectively. As $y_1 = y(r_1) + c_1$, $y_2 = y(r_2) + c_2$, then $c_1 - c_2 = 2\pi k$. Hence, $\bar{y}(r) - \tilde{y}(r) = 2\pi k$ for any r. Therefore, the surface $\psi(l_1)$ coincides with the surface $\psi(l_2)$, i.e. they form the same fiber. Evidently, every point within K^3 has a neighborhood covered with the fibers $\psi(l)$. The boundary torus is also the fiber of foliation just obtained.

The sphere S^3 can be represented as a union of two toroidal bodies with a common boundary, namely the torus T^2. Within each toroidal body we construct the foliation as above. Gluing the toroidal bodies along their common boundary T^2, we get S^3 endowed with the Reeb foliation.

Observe that in this case we have two closed curves which are orthogonal to the fibers, namely the axial curves of the toroidal bodies.

Later, the theory of foliations was developed in topological and geometrical directions. Novikov [73] proved that every codimension-one foliation on S^3 has a compact leaf. Thurston [100] proved the existence of p-dimensional foliations on any opened manifold. These foliations exist on closed manifolds if there exists a field of p-dimensional tangent spaces. Reviews of topological results from the theory of foliations may be found in the work of Fuks [101], Tamura [71] and Lawson [102].

A series of interesting results was obtained for foliations in Riemannian spaces where the leaves were totally geodesic, minimal, umbilical and so on. It was proved that the codimension of a totally geodesic foliation on a complete manifold of positive curvature is more than 1. Moreover, the Frankel theorem is known: in a compact manifold with positive curvature, two compact totally geodesic submanifolds whose summed dimensions are not less than the dimension of the manifold do not have an empty intersection. So the dimension of a totally geodesic foliation with compact leaves on a compact manifold with a positive curvature is less than half the manifold's dimension.

Ferus [103] proved the following interesting theorem in which the completeness of the manifold is not necessary: if in some domain of the manifold $M^{n+\nu}$ there exists a totally geodesic foliation with dimension ν of class C^2 and with *complete leaves* and a mixed curvature of $M^{n+\nu}$ tangential to a leaf of vector X and orthogonal to a leaf of vector Y,

$$K(X, Y) = const > 0,$$

so $\nu < \rho(n)$, where $(\rho(n) - 1)$ is the maximal number of linearly independent vector fields on S^{n-1}.

Rovenskii [104] showed that the condition $K(X, Y) = const$ is essential and he generalized this theorem for some class of foliations, which he called conformal. Papers [105]–[113] give various results on foliations with geometrical conditions.

There is a review of geometrical results on foliations in the work of Tondeur [114].

2.17 The Godbillon–Vey Invariant for the Foliation on a Manifold

Let us be given a C^2-regular foliation of codimension 1 in a differentiable manifold M^n of class C^∞. The foliation is called *transversally orientable* if there is a smooth vector field on M^n transversal (i.e. not tangent) to the foliation leaf through the initial point. Consider the transversally orientable foliation. Then in every neighborhood of a point the foliation leaves can be defined by the Pfaff equation $\omega_1 = 0$. The form ω_1 must satisfy the integrability condition $\omega_1 \wedge d\omega_1 = 0$. This means that $d\omega_1$ is divisible by ω_1, i.e. there is a form ω_2 such that

$$d\omega_1 = \omega_1 \wedge \omega_2. \tag{1}$$

The form ω_2 is not uniquely defined. If we differentiate (1) externally then we obtain

$$\omega_1 \wedge d\omega_2 = 0. \tag{2}$$

Therefore, $d\omega_2$ is also divisible by ω_1. There is a form ω_3 such that

$$d\omega_2 = \omega_1 \wedge \omega_3. \tag{3}$$

If we differentiate this new relation then we obtain

$$\omega_1 \wedge (\omega_2 \wedge \omega_3 - d\omega_3) = 0.$$

Consider the third-order form $\Omega = \omega_1 \wedge \omega_2 \wedge \omega_3$ which was introduced by Godbillon and Vey in [68]. Using (3), *the form Ω can be expressed as*

$$\Omega = -\omega_2 \wedge d\omega_2. \tag{4}$$

For the case of $n = 3$ the form Ω can be expressed in terms of the value of the non-holonomicity of some vector field. Since ω_2 is linear, it can be written as $\omega_2 = P\,dx^1 + Q\,dx^2 + R\,dx^3$. If we introduce the vector field $\nu = \{P, Q, R\}$ then $\Omega = -(\nu, \operatorname{curl} \nu) dx^1 \wedge dx^2 \wedge dx^3$.

The following theorem holds:

Theorem *The form Ω is closed and its cohomology class in $H^3(M, R)$ depends only on the foliation F*

Consider the exterior differential of Ω:

$$d\Omega = -d\omega_2 \wedge d\omega_2 = -\omega_1 \wedge \omega_3 \wedge \omega_1 \wedge \omega_3.$$

Since ω_1 is a form of odd order, $\omega_1 \wedge \omega_1 = 0$. Hence $d\Omega = 0$.

Suppose that ω_2' is any other form satisfying $d\omega_1 = \omega_1 \wedge \omega_2'$. Let us show that the Godbillon–Vey form constructed by means of ω_2' differs from the previous form by some differential form. We have

$$d\omega_1 = \omega_1 \wedge \omega_2' = \omega_1 \wedge \omega_2.$$

From this we conclude that $\omega_1 \wedge (\omega_2' - \omega_2) = 0$, which means that $\omega_2' - \omega_2$ is proportional to ω_1, i.e. there is a function f such that $\omega_2' - \omega_2 = f\omega_1$.

We have

$$\begin{aligned} \omega_2' \wedge d\omega_2' = {} & \omega_2 \wedge d\omega_2 + f\omega_1 \wedge d\omega_2 + \omega_2 \wedge df \wedge \omega_1 \\ & + f\omega_1 \wedge df \wedge \omega_1 + f\omega_2 \wedge d\omega_1 + f^2\omega_1 \wedge d\omega_1. \end{aligned}$$

Since ω_1 corresponds to the foliation, $\omega_1 \wedge d\omega_1 = 0$. Also, $\omega_1 \wedge d\omega_2 = 0$, $\omega_1 \wedge df \wedge \omega_1 = 0$. Then

$$\begin{aligned} -d(f\omega_2 \wedge \omega_1) &= -df \wedge \omega_2 \wedge \omega_1 - f d\omega_2 \wedge \omega_1 + f\omega_2 \wedge d\omega_1 \\ &= \omega_2 \wedge df \wedge \omega_1 + f\omega_2 \wedge d\omega_1. \end{aligned}$$

Hence

$$\omega_2' \wedge d\omega_2' = \omega_2 \wedge d\omega_2 - d(f\omega_2 \wedge \omega_1).$$

So, the forms $-\omega_2 \wedge d\omega_2$ and $-\omega_2' \wedge d\omega_2'$ belong to the same cohomology class of $H^3(M, R)$. This class is called *the Godbillon–Vey characteristic class*. Denote it by $[\Omega](F)$.

Now we are going to establish the independence of $[\Omega](F)$ of the concrete choice of ω_1, i.e. we show that it depends only on the foliation F itself. The form ω_1 is definite to the scalar multiplier. Let $\omega_1' = \lambda\omega_1$ be the form which gives the same foliation F, where λ is a regular function. Set $d\omega_1' = \omega_1' \wedge \omega_2'$. We have

$$\begin{aligned} d\omega_1' &= d\lambda\omega_1 = d\lambda \wedge \omega_1 + \lambda\, d\omega_1 = d\lambda \wedge \omega_1 + \lambda\omega_1 \wedge \omega_2 \\ &= \lambda\omega_1 \wedge (\omega_2 - d \ln \lambda) = \omega_1' \wedge \omega_2'. \end{aligned}$$

Hence, there is a function f such that

$$\omega_2' = \omega_2 - d \ln \lambda + f\omega_1.$$

We have

$$\begin{aligned} \omega_2' \wedge d\omega_2' = \omega_2 \wedge d\omega_2 &- d \ln \lambda \wedge d\omega_2 + f\omega_1 \wedge d\omega_2 \\ &+ \omega_2 \wedge df \wedge \omega_1 - d \ln \lambda \wedge df \wedge \omega_1 + f\omega_1 \wedge df \wedge \omega_1 \\ &+ f\omega_2 \wedge d\omega_1 - fd \ln \lambda \wedge d\omega_1 + f^2\omega_1 \wedge d\omega_1. \end{aligned}$$

Many of the terms in the expression above vanish by virtue of the equalities

$$\omega_2 \wedge d\omega_1 = 0, \qquad \omega_1 \wedge d\omega_1 = 0, \qquad \omega_1 \wedge df \wedge \omega_1 = 0,$$
$$\omega_1 \wedge d\omega_2 = 0, \qquad \omega_2 \wedge d\omega_1 = 0.$$

Hence, we finally obtain

$$\omega_2' \wedge d\omega_2' = \omega_2 \wedge d\omega_2 - d(\ln \lambda\, d\omega_2 + f\omega_2 \wedge \omega_1 - d \ln \lambda \wedge \omega_1).$$

This equality means that $-\omega_2' \wedge d\omega_2'$ and $-\omega_2 \wedge d\omega_2$ differ from each other by the differential of some second-order form. Thus, the cohomology class $[\Omega](F)$ depends on the specific foliation F.

We present some properties of $[\Omega](F)$:

(1) Let us be given a foliation F in M. Let $f\colon V \to M$ be a differentiable mapping of a manifold V onto M. Then the inverse images of fibers of the foliation F define a foliation in V. Denote this foliation by $f^*(F)$. The mapping f induces a mapping

f^* of the cohomology group $H^3(M, R)$ into $H^3(V, R)$:

$$f^*: H^3(M, R) \to H^3(V, R).$$

Then *the Godbillon–Vey class* $[\Omega](F)$ *of the foliation* F *goes into the Godbillon–Vey class of induced foliation* $f^*(F)$ *under the mapping* f^*.

$$f^*([\Omega](F)) = [\Omega](f^*(F)).$$

This property of Godbillon–Vey classes has been proved in [98] for the case of C^2-regular foliations and C^1-diffeomorphisms.

(2) *If* F *and* F' *are the foliations of codimension 1 in the same manifold* M *which are differentiable and homotopic to each other then their Godbillon–Vey classes coincide.*

(3) *If the foliation* F *is given in a three-dimensional manifold* M^3 *then the integral of* Ω *gives some number* γ, namely

$$\gamma = \int_{M^3} \Omega = -\int_{M^3} \omega_2 \wedge d\omega_2.$$

This number is uniquely defined by the foliation and is called *the Godbillon-Vey number*. This number is not necessarily an integer.

(4) *Let us settle the relation of* Ω *to both the curvatures of lines which are orthogonal to the fibers and the second fundamental forms of the fibers.* Denote the lines orthogonal to the fibers by l. Let $k_1, \dots, k_{n-1}$ be the curvatures of those lines. Suppose that $\tau_1, \dots, \tau_n$ is the natural orthonormal frame of l, where $\tau_i = \{\tau_{ij}\}$. Let ζ_i be a linear form generated by τ_i. The foliation F is given by the equation $\zeta_1 = 0$. In Section 2.11 we proved that the exterior differential of ζ_1 can be represented as the exterior product of ζ_1 and $k_1\zeta_2$, i.e.

$$d\zeta_1 = \zeta_1 \wedge k_1\zeta_2.$$

Therefore, the Godbillon–Vey class for the foliation F is represented by the form

$$\Omega = -k_1^2\zeta_2 \wedge d\zeta_2.$$

Let us express the form $d\zeta_2$ in terms of an exterior product of the forms ζ_i. Let $\alpha, \beta \geq 3$. Find the value of $d\zeta_2$ on τ_α, τ_β. Using (5) from Section 2.11, we have

$$d\zeta_2(\tau_\alpha, \tau_\beta) = (\tau_2, \nabla_{\tau_\alpha}\tau_\beta - \nabla_{\tau_\beta}\tau_\alpha).$$

Consider the system of two Pfaff equations: $\zeta_1 = 0$, $\zeta_2 = 0$. By virtue of (1) and (3), the differentials of ζ_1 and ζ_2 can be expressed as

$$d\zeta_1 = k_1\zeta_1 \wedge \zeta_2, \quad d\zeta_2 = \zeta_1 \wedge \varphi,$$

where φ is some linear form. *Since* $d\zeta_1 = 0$ *and* $d\zeta_2 = 0$ *by virtue of* $\zeta_1 = 0$, *the system* $\zeta_1 = 0, \zeta_2 = 0$ *is totally integrable and, as a consequence, defines a foliation of codimension* 2 *such that each fiber of the new foliation is in some fiber of the given foliation F*. Therefore, the vector field τ_2 is orthogonal to the Poisson bracket of any vector fields τ_α, τ_β for $\alpha, \beta \geq 3$. Hence

$$d\zeta_2(\tau_\alpha, \tau_\beta) = 0.$$

Next we consider the value of this form on τ_1, τ_α. We have the Frenet equations

$$\begin{aligned}
\nabla_{\tau_1}\tau_1 &= k_1\tau_2,\\
\nabla_{\tau_1}\tau_2 &= -k_1\tau_1 + k_2\tau_3,\\
&\dots\\
\nabla_{\tau_1}\tau_i &= -k_{i-1}\tau_{i-1} + k_i\tau_{i+1}.
\end{aligned}$$

Taking them into account, we are able to write

$$\begin{aligned}
d\zeta_2(\tau_1, \tau_\alpha) &= (\tau_2, \nabla_{\tau_1}\tau_\alpha - \nabla_{\tau_\alpha}\tau_1)\\
&= (\tau_2, -k_{\alpha-1}\tau_{\alpha-1} + k_\alpha\tau_{\alpha+1} - \nabla_{\tau_\alpha}\tau_1).
\end{aligned}$$

Denote the second fundamental form of the fiber of the foliation F by $l(x, y)$. Then

$$d\zeta_2(\tau_1, \tau_\alpha) = -k_2\delta_3^\alpha - l(\tau_2, \tau_\alpha), \quad \alpha \geq 2,$$

where δ_3^α is the Kronecker symbol. Thus, $d\zeta_2$ can be expressed in terms of ζ_i as

$$d\zeta_2 = -k_2\zeta_1 \wedge \zeta_3 - \sum_{\alpha=3}^{n} l(\tau_2, \tau_\alpha)\zeta_1 \wedge \zeta_\alpha + c\zeta_1 \wedge \zeta_2,$$

where c is some unimportant coefficient. So, we find the following *expression for the Godbillon–Vey form*:

$$\Omega = -k_1^2\zeta_2 \wedge d\zeta_2 = k_1^2\left(k_2\zeta_2 \wedge \zeta_1 \wedge \zeta_3 + \sum_{\alpha=3}^{n} l(\tau_2, \tau_\alpha)\zeta_2 \wedge \zeta_1 \wedge \zeta_\alpha\right).$$

This expression was established in [69].

2.18 The Expression for the Hopf Invariant in Terms of the Integral of the Field Non-Holonomicity Value

Let $f\colon S^3 \to S^2$ be a C^2-regular mapping of spheres. We are going to find the expression for the Hopf invariant γ of a mapping f in terms of the integrated non-holonomicity value of some vector field given on S^3. Let u^1, u^2 be the local coordinates in S^2 and $ds^2 = g_{ij}\,du^i\,du^j$ be the first fundamental form of S^2 with respect to these coordinates, $g = \det \|g_{ij}\|$. Let x^1, x^2, x^3 be the local coordinates in S^3. The mapping f can be represented with the functions

$$u^i = u^i(x^1, x^2, x^3), \quad i = 1, 2, 3.$$

Set

$$I_{\alpha\beta} = \begin{vmatrix} u^1_{x^\alpha} & u^1_{x^\beta} \\ u^2_{x^\alpha} & u^2_{x^\beta} \end{vmatrix}.$$

Let us introduce in S^3 a second-order skew-symmetric tensor $a_{\alpha\beta} = \frac{1}{4\pi}\sqrt{g}\, I_{\alpha\beta}$. This tensor generates the second-order exterior form, namely

$$\omega = a_{\alpha\beta}\, dx^\alpha \wedge dx^\beta.$$

Show that the exterior differential of this form is zero: $d\omega = 0$. Indeed, we have

$$d\omega = \left\{\frac{\partial a_{12}}{\partial x^3} + \frac{\partial a_{23}}{\partial x^1} + \frac{\partial a_{31}}{\partial x^2}\right\} dx^1 \wedge dx^2 \wedge dx^3.$$

Since $\sqrt{g}$ is a function of u^i which are the functions of x^α on the other hand,

$$\frac{\partial a_{ij}}{\partial x^\alpha} = \frac{1}{4\pi}\left\{\frac{\partial\sqrt{g}}{\partial u^k}\frac{\partial u^k}{\partial x^\alpha} I_{ij} + \sqrt{g}\frac{\partial I_{ij}}{\partial x^\alpha}\right\}.$$

It is easy to find the following equalities:

$$\frac{\partial I_{23}}{\partial x^1} + \frac{\partial I_{31}}{\partial x^2} + \frac{\partial I_{12}}{\partial x^3} = 0, \quad \frac{\partial u^i}{\partial x^1} I_{23} + \frac{\partial u^i}{\partial x^2} I_{31} + \frac{\partial u^i}{\partial x^3} I_{12} = \begin{vmatrix} u^i_{x^1} & u^i_{x^2} & u^i_{x^3} \\ u^1_{x^1} & u^1_{x^2} & u^1_{x^3} \\ u^2_{x^1} & u^2_{x^2} & u^2_{x^3} \end{vmatrix} = 0.$$

These equalities imply $d\omega = 0$, i.e. the form ω is closed. Since the cohomology group $H^2(S^3, R)$ is trivial, then by the de Rham theorem there is a global linear form $\omega_1 = b_i\, dx^i$ on S^3 such that $\omega = d\omega_1$. Whitehead [66] stated the following assertion.

Theorem *The Hopf invariant γ of a mapping $f: S^3 \to S^2$ is equal to the integrated non-holonomicity value of the field $\mathbf{b} = \{b_i\}$*

$$\gamma = \int\limits_{S^3} \omega_1 \wedge d\omega_1 = \int\limits_{S^3} (\mathbf{b}, \operatorname{curl}\mathbf{b})\, dV, \tag{1}$$

where dV is a volume element of S^3.

Let us divide S^2 into two-dimensional cells E_i^2 which are so small that the mapping f can be approximated with the normal mapping. Further on we assume that f is already normal within each cell E_i^2.

Let $M_q = f^{-1}(q)$. As f is normal, the set M_q consists of one or several closed curves. Let us orient each curve of M_q as follows. Let p be a point in M_q. Select the basis $\mathbf{e}_1, \mathbf{e}_2, \mathbf{e}_3$ at p which produces a positive orientation of S^3 and such that $\mathbf{e}_3$ is tangent to M_q, while the directions of $\mathbf{e}_1$ and $\mathbf{e}_2$ go into the directions of $\mathbf{e}_1^*$ and $\mathbf{e}_2^*$ under the induced mapping of tangent spaces f^*: $T_p(S^3) \to T_q(S^2)$, such that $(\mathbf{e}_1^*, \mathbf{e}_2^*)$ produce a positive orientation of S^2. Then we say that $\mathbf{e}_3$ defines a positive orientation in M_q.

Let q_1, q_2 be points in E_l^2. Connect them by a smooth curve l in E_l^2. Then $M_l = f^{-1}(l)$ is a surface with the oriented boundary $\partial M_l = M_{q_2} - M_{q_1}$. The whole surface is formed by M_q, where $q \in l$. Show that

$$\int\limits_{M_{q_1}} \omega_1 = \int\limits_{M_{q_2}} \omega_1.$$

Indeed, by Stokes' theorem we have

$$\int\limits_{M_l} d\omega_1 = \int\limits_{M_{q_1}} \omega_1 - \int\limits_{M_{q_2}} \omega_1. \tag{2}$$

Let v^1, v^2 be local coordinates in M_l. Set

$$T^{\alpha\beta} = \begin{vmatrix} \dfrac{\partial x^\alpha}{\partial v^1} & \dfrac{\partial x^\alpha}{\partial v^2} \\ \dfrac{\partial x^\beta}{\partial v^1} & \dfrac{\partial x^\beta}{\partial v^2} \end{vmatrix}.$$

The coordinates u^1, u^2 of a point in S^2 are the functions of v^1, v^2 along M_l. Then

$$I\left(\frac{u^1, u^2}{v^1, v^2}\right) = \begin{vmatrix} u^1_{x^\alpha} x^\alpha_{v^1} & u^1_{x^\beta} x^\beta_{v^2} \\ u^2_{x^\alpha} x^\alpha_{v^1} & u^2_{x^\beta} x^\beta_{v^2} \end{vmatrix} = x^\alpha_{v^1} x^\beta_{v^2} I_{\alpha\beta} = \frac{1}{2} T^{\alpha\beta} I_{\alpha\beta}.$$

Since the surface M_l goes into the curve under the mapping f, $T\left(\frac{u^1, u^2}{v^1, v^2}\right) = 0$. Consider the integral on the left of (2):

$$\begin{aligned} \int\limits_{M_l} d\omega_1 &= \int\limits_{M_l} \omega = \frac{1}{4\pi} \int\limits_{M_l} \sqrt{g} I_{\alpha\beta} dx^\alpha \wedge dx^\beta \\ &= \frac{1}{4\pi} \int\limits_{M_l} \frac{\sqrt{g}}{2} I_{\alpha\beta} T^{\alpha\beta} dv^1 \wedge dv^2 \\ &= \frac{1}{8\pi} \int\limits_{M_l} \sqrt{g} I\left(\frac{u^1, u^2}{v^1, v^2}\right) dv^1 \wedge dv^2 = 0. \end{aligned}$$

Equality (1) is proved. Thus, the integral of ω_1 over the inverse image of any point q is the same number — some invariant of the mapping f. Let us show that this number coincides with the Hopf invariant γ. Let M_q be a base of a piecewise regular surface F and ξ, η the local coordinates in F. Then by Stokes' theorem we have

$$\int\limits_{M_q} \omega_1 = \int\limits_F d\omega_1 = \frac{1}{4\pi} \int\limits_F \sqrt{g} I\left(\frac{u^1, u^2}{\xi, \eta}\right) d\xi \wedge d\eta.$$

As the boundary of F goes into the single point q, then $f(F)$ covers the sphere S^2 some integer number of times (more precisely, the total oriented area of the mapping

f image is a multiple of 4π). This number is called the Hopf invariant of mapping f. So,

$$\int_{M_q} \omega_1 = \gamma.$$

Consider a neighborhood T of the curve M_q in S^3. It possible to represent T as a topological product of M_q and some surface Φ for which the mapping f maps onto E_i^2, i.e. $T = M_q \times \Phi$. Let u^3 be the parameter in M_q ranging from 0 to 2π and u^1, u^2 be the coordinates in Φ induced from E_i^2. Then every point p in T has the curvilinear coordinates u^1, u^2, u^3 and the mapping f translates the point p into the point q with coordinates u^1, u^2. With respect to coordinates u^1, u^2, u^3 the form ω can be written as

$$\omega = \frac{\sqrt{g}}{4\pi}\begin{vmatrix} u^1_{u^\alpha} & u^1_{u^\beta} \\ u^2_{u^\alpha} & u^2_{u^\beta} \end{vmatrix} du^\alpha \wedge du^\beta = \frac{\sqrt{g}}{4\pi}\, du^1 \wedge du^2.$$

Let $\omega_1 = c_i\, du^i$, where c_i are some functions of u^1, u^2, u^3. Consider the exterior product

$$\omega_1 \wedge d\omega_1 = \omega_1 \wedge \omega = \frac{c_3\sqrt{g}}{4\pi}\, du^1 \wedge du^2 \wedge du^3.$$

Note that $\sqrt{g}$ does not depend on u^3. Represent the integral of $\omega_1 \wedge d\omega_1$ over the domain T as any iterated one as follows:

$$\int_T \omega_1 \wedge d\omega_1 = \int_{E_i^2} \frac{\sqrt{g}}{4\pi}\, du^1 \wedge du^2 \int_0^{2\pi} c_3(u^1, u^2, u^3)\, du^3 = \int_{E_i^2} \frac{\sqrt{g} du^1 du^2}{4\pi} \int_{M_q} \omega_1.$$

Since the integral over M_q of ω_1 does not depend on q,

$$\int_T \omega_1 \wedge d\omega_1 = \frac{\gamma S(E_i^2)}{4\pi},$$

where $S(E_i^2)$ stands for the area of E_i^2 in S^2. If we sum over all domains E_i^2 then we obtain (1). The theorem is proved.

S. Novikov [74] obtained some many-dimensional generalizations of the Whitehead formula.

2.19 Vector Fields Tangent to Spheres

Consider the vector fields in E^n which are tangent to a sphere S^{n-1} at each point $x \in S^{n-1}$ from the families of spheres with a common center. It is well known that an even-dimensional sphere admits no regular tangent vector fields. Therefore, we consider the vector fields on the odd-dimensional spheres.

The vector fields $\mathbf{n}_1(x), \ldots, \mathbf{n}_k(x)$ on a sphere S^{n-1} are called *linearly independent* if they are linearly independent at each point $x \in S^{n-1}$. The Hurwitz–Radon–Eckmann theorem [93] on the possible number of linearly independent vector fields on S^n is known. James formulated this result as follows: represent n in the form of a product of an odd number and a power of 2, i.e. $n = (2a+1)2^b$. Divide b by 4 with the remainder, i.e. represent it in the form $b = 4d + c$, where $0 \le c \le 3$. Set $\rho(n) = 2^c + 8d$. Then there are $\rho(n) - 1$ linearly independent vector fields on the sphere S^{n-1}. Adams [94] proved that this number is maximally possible.

We denote the number of linearly independent vector fields on S^{n-1} by k. Below we present a table of k for low-dimensional spheres:

$n-1$	1	3	5	7	9	11	13	15	17	19	21	23	25	27	29
k	1	3	1	7	1	3	1	8	1	3	1	7	1	3	1

We shall construct the regular vector fields on low-dimensional spheres in explicit form, avoiding heavy theory. This also gives us a rich set of non-holonomic fields in E^n.

It is easy to find a single vector field on the odd-dimensional sphere. Suppose that S^{n-1} is given as a set of points in E^n whose coordinates satisfy the equation

$$x_1^2 + \ldots + x_n^2 = \rho^2,$$

where ρ is a radius of S^{n-1}.

Since n is odd, the coordinates can be gathered into pairs, for instance, in consequent order. We define the vector field $\mathbf{a} = (-x_2, x_1, -x_4, x_3, \ldots, -x_n, x_{n-1})$ at the point of a sphere having coordinates $x_1, \ldots, x_n$.

Let us construct three mutually orthogonal vector fields on S^3. Denote a position vector of a point in S^3 by $\mathbf{r} = (x_1, \ldots, x_4)$. Split the coordinates into two pairs: (x_1, x_2), (x_3, x_4). Perform the permutation of the form $\begin{pmatrix} x_i & x_j \\ -x_j & x_i \end{pmatrix}$ over each pair. The splitting under consideration gives the vector field $\mathbf{a} = (-x_2, x_1, -x_3, x_4)$. Next we take the splitting into pairs (x_1, x_3), (x_2, x_4) and construct the vector field $\mathbf{b} = (-x_3, x_4, x_1, -x_2)$ in an analogous way to the construction of the field $\mathbf{a}$. Note, however, that the field $(-x_3, -x_4, x_1, x_2)$ can also be constructed by the procedure above, but it is not orthogonal to $\mathbf{a}$. Taking the pairs (x_1, x_4), (x_2, x_3), we find the vector field $\mathbf{c} = (-x_4, -x_3, x_2, x_1)$. Write the matrix formed by the components of $\mathbf{r}, \mathbf{a}, \mathbf{b}, \mathbf{c}$:

r	x_1	x_2	x_3	x_4
a	$-x_2$	x_1	$-x_4$	x_3
b	$-x_3$	x_4	x_1	$-x_2$
c	$-x_4$	$-x_3$	x_2	x_1

It is easy to see that the constructed fields are tangent to the sphere S^3 centered at the origin. They are mutually orthogonal and of unit length. We shall say that each of the vectors $\mathbf{a}, \mathbf{b}, \mathbf{c}$ defines some permutations over coordinates, say A, B, C.

For instance, the permutation A applied to the quadruple $(\xi_1, \xi_2, \xi_3, \xi_4)$ gives $(-\xi_2, \xi_1, -\xi_4, \xi_3)$.

Let us construct the vector fields on S^7. We split eight coordinates into two groups of four coordinates: (x_1, x_2, x_3, x_4), (x_5, x_6, x_7, x_8). For each of these groups we can construct the fields analogous to those on S^3. We obtain four of the vector fields on S^3 by joining three-dimensional sphere fields, i.e. we define them as follows:

$$\mathbf{a} = (\mathbf{a}_1, \mathbf{a}_2) = (-x_2, x_1, -x_4, x_3, -x_6, x_5, -x_8, x_7),$$
$$\mathbf{b} = (\mathbf{b}_1, -\mathbf{b}_2) = (-x_3, x_4, x_1, -x_2, x_7, -x_8, -x_5, x_6),$$
$$\mathbf{c} = (\mathbf{c}_1, \mathbf{c}_2) = (-x_4, -x_3, -x_2, x_1, -x_8, -x_7, x_6, x_5),$$

where $\mathbf{a}_i, \mathbf{b}_i, \mathbf{c}_i$ are the vectors on S^3, constructed as above involving the coordinates of the i-th group. Evidently, these fields are mutually orthogonal.

Consider the splitting into the following quadruples: (x_1, x_2, x_5, x_6), (x_3, x_4, x_7, x_8). Perform the B permutation over each of them:

$$\begin{pmatrix} x_1, x_2, x_5, x_6 \\ -x_5, x_6, x_1, -x_2 \end{pmatrix}, \quad \begin{pmatrix} x_3, x_4, x_7, x_8 \\ -x_7, x_8, x_3, -x_4 \end{pmatrix}.$$

These permutations define the vector field

$$\mathbf{d} = (-x_5, x_6, -x_7, x_8, x_1, -x_2, x_3, -x_4).$$

In an analogous way, the permutation C over the same groups defines

$$\mathbf{e} = (-x_6, -x_5, x_8, x_7, x_2, x_1, -x_4, -x_3).$$

One can check that the constructed fields are mutually orthogonal. Now consider the groups (x_1, x_2, x_7, x_8) and (x_3, x_4, x_5, x_6). Perform the permutations B, C over each of them. This gives two more vector fields:

$$\mathbf{f} = (-x_7, x_8, x_5, -x_6, -x_3, x_4, x_1, -x_2),$$
$$\mathbf{g} = (-x_8, -x_7, -x_6, -x_5, x_4, x_3, x_2, x_1).$$

Write the matrix formed with the components of $\mathbf{a}, \dots, \mathbf{g}$ on S^7 and a position vector $\mathbf{r}$:

r	x_1	x_2	x_3	x_4	x_5	x_6	x_7	x_8
a	$-x_2$	x_1	$-x_4$	x_3	$-x_6$	x_5	$-x_8$	x_7
b	$-x_3$	x_4	x_1	$-x_2$	x_7	$-x_8$	$-x_5$	x_6
c	$-x_4$	$-x_3$	x_2	x_1	$-x_8$	$-x_7$	x_6	x_5
d	$-x_5$	x_6	$-x_7$	x_8	x_1	$-x_2$	x_3	$-x_4$
e	$-x_6$	$-x_5$	x_8	x_7	x_2	x_1	$-x_4$	$-x_3$
f	$-x_7$	x_8	x_5	$-x_6$	$-x_3$	x_4	x_1	$-x_2$
g	$-x_8$	$-x_7$	$-x_6$	$-x_5$	x_4	x_3	x_2	x_1

The components of this matrix satisfy $a_{ij} = -a_{ji}$ for $i \neq j$. It is easy to check that all constructed vector fields on S^7 are mutually orthogonal.

As we know from the table at the beginning of this section, the sphere S^{11} only admits three linearly independent vector fields. To find them, we split twelve coordinates x_i into three groups of four and perform the permutations A, B, C over each of them. Write the corresponding matrix of components of vector fields on S^{11}:

r	x_1	x_2	x_3	x_4	x_5	x_6	x_7	x_8	x_9	x_{10}	x_{11}	x_{12}
a	$-x_2$	x_1	$-x_4$	x_3	$-x_6$	x_5	$-x_8$	x_7	$-x_{10}$	x_9	$-x_{12}$	x_{11}
b	$-x_3$	x_4	x_1	$-x_2$	x_7	$-x_8$	$-x_5$	x_6	$-x_{11}$	x_{12}	x_9	$-x_{10}$
c	$-x_4$	$-x_3$	x_2	x_1	$-x_8$	$-x_7$	x_6	x_5	$-x_{12}$	$-x_{11}$	x_{10}	x_9

Evidently, a more general statement holds: *if the sphere S^{n-1} admits k linearly independent vector fields then the sphere S^{nq-1} admits not less than k vector fields.* As $23 = 8 \cdot 3 - 1$, S^{23} admits seven vector fields which are the three times repetition of the seven vector fields on S^7.

Let us construct eight vector fields on S^{15}. Since we can split the sixteen coordinates into two groups by eight coordinates, namely $(x_1, \dots, x_8)$ and $(x_9, \dots, x_{16})$, and we have already constructed seven vector fields on S^7, by double repetition of these fields we obtain seven vector fields on S^{15}. If $a_i, b_i, \dots, g_i$ $i = 1, 2$ are the corresponding vector fields on S^7 generated by the splitting mentioned above then we represent the seven vector fields on S^{15} as $\mathbf{a} = (a_1, a_2)$, $\mathbf{b} = (b_1, b_2)$, ..., $\mathbf{g} = (g_1, g_2)$.

Now construct one more vector field on S^{15} which is orthogonal to all the other fields. Consider one more splitting, namely $(x_1, x_2, x_3, x_4, x_9, x_{10}, x_{11}, x_{12})$ and $(x_5, x_6, x_7, x_8, x_{13}, x_{14}, x_{15}, x_{16})$. Construct the vector $\mathbf{h}$ formed as (d_1, d_2) with respect to the groups of coordinates given, i.e. the vector obtained by means of the D-type permutation (see the structure of the field $\mathbf{d}$ on S^7) $\mathbf{h} = (-x_9, x_{10}, -x_{11}, x_{12}, -x_{13}, x_{14}, -x_{15}, x_{16}, x_1, -x_2, x_3, -x_4, x_5, -x_6, x_7, -x_8)$. However, a direct check shows that $\mathbf{h}$ is not orthogonal to $\mathbf{b}, \mathbf{d}, \mathbf{f}$. To satisfy the condition of orthogonality we ought to change the signs of the last eight components of these vectors, i.e. we ought to set $\mathbf{b} = (b_1, -b_2)$, $\mathbf{d} = (d_1, -d_2)$, $\mathbf{f} = (f_1, -f_2)$. Present the table formed with components of mutually orthogonal vector fields on S^{15} and $\mathbf{r}$:

r	x_1	x_2	x_3	x_4	x_5	x_6	x_7	x_8	x_9	x_{10}	x_{11}	x_{12}	x_{13}	x_{14}	x_{15}	x_{16}
a	$-x_2$	x_1	$-x_4$	x_3	$-x_6$	x_5	$-x_8$	x_7	$-x_{10}$	x_9	$-x_{12}$	x_{11}	$-x_{14}$	x_{13}	$-x_{16}$	x_{15}
b	$-x_3$	x_4	x_1	$-x_2$	x_7	$-x_8$	$-x_5$	x_6	x_{11}	$-x_{12}$	$-x_9$	x_{10}	$-x_{15}$	x_{16}	x_{13}	$-x_{14}$
c	$-x_4$	$-x_3$	x_2	x_1	$-x_8$	$-x_7$	x_6	x_5	$-x_{12}$	$-x_{11}$	x_{10}	x_9	$-x_{16}$	$-x_{15}$	x_{14}	x_{13}
d	$-x_5$	x_6	$-x_7$	x_8	x_1	$-x_2$	x_3	$-x_4$	x_{13}	$-x_{14}$	x_{15}	$-x_{16}$	$-x_9$	x_{10}	$-x_{11}$	x_{12}
e	$-x_6$	$-x_5$	x_8	x_7	x_2	x_1	$-x_4$	$-x_3$	$-x_{14}$	$-x_{13}$	x_{16}	x_{15}	x_{10}	x_9	$-x_{12}$	$-x_{11}$
f	$-x_7$	x_8	x_5	$-x_6$	$-x_3$	x_4	x_1	$-x_2$	x_{15}	$-x_{16}$	$-x_{13}$	x_{14}	x_{11}	$-x_{12}$	$-x_9$	x_{10}
g	$-x_8$	$-x_7$	$-x_6$	$-x_5$	x_4	x_3	x_2	x_1	$-x_{16}$	$-x_{15}$	$-x_{14}$	$-x_{13}$	x_{12}	x_{11}	x_{10}	x_9
h	$-x_9$	x_{10}	$-x_{11}$	x_{12}	$-x_{13}$	x_{14}	$-x_{15}$	x_{16}	x_1	$-x_2$	x_3	$-x_4$	x_5	$-x_6$	x_7	$-x_8$

We can construct the vector field $\mathbf{i}$ by means of a permutation of E types over the second splitting (see the structure of the field $\mathbf{e}$ on S^7):

$$\mathbf{i} = (-x_{10}, -x_9, x_{12}, x_{11}, -x_{14}, -x_{13}, x_{16}, x_{15}, x_2, x_1, -x_4, -x_3, x_6, x_5, -x_8, -x_7).$$

A direct check shows that $\mathbf{i}$ is orthogonal to all the fields except $\mathbf{d}$: $(\mathbf{i}, \mathbf{d}) \neq 0$. Thus, the set of mutually orthogonal vector fields is not uniquely defined. Instead of $\mathbf{d}$ we can take $\mathbf{i}$.

Consider the geometrical properties of the fields just constructed. Each of these fields is non-holonomic. Let us show, for instance, that $\mathbf{a}$ is a field in E^n which is non-holonomic, i.e. there is no family of the field $\mathbf{a}$ orthogonal to hypersurfaces. Consider the Pfaff equation which corresponds to the field $\mathbf{a}$:

$$\omega = -x_2\,dx_1 + x_1\,dx_2 + \cdots - x_n\,dx_{n-1} + x_{n-1}\,dx_n = 0. \tag{1}$$

The exterior differential of this form is

$$d\omega = 2\{dx_1 \wedge dx_2 + \cdots + dx_{n-1} \wedge dx_n\}. \tag{2}$$

Exclude one of the differentials x_i from (1), say dx_1, and substitute into (2); we obtain

$$d\omega = 2\left\{\sum_{i=3}^{n} \varphi_i\,dx_i \wedge dx_2 + dx_3 \wedge dx_4 + \cdots + dx_{n-1} \wedge dx_n\right\},$$

where φ_i are some functions of $x_2, \ldots, x_n$. The exterior differential $d\omega$ is not zero identically by virtue of $\omega = 0$. So, the field $\mathbf{a}$ is non-holonomic.

The streamlines of every field under consideration are great circles of spheres. Consider, for instance, the streamlines of the field $\mathbf{a}$ on S^3. The derivatives of the position vector of the streamline satisfy

$$\frac{dx_1}{ds} = -x_2, \quad \frac{dx_2}{ds} = x_1, \quad \frac{dx_3}{ds} = -x_4, \quad \frac{dx_4}{ds} = x_3,$$

i.e. $d\mathbf{r}/ds = \mathbf{a}$. Hence,

$$\frac{d^2x_1}{ds^2} = -x_1, \quad \cdots \quad , \frac{d^2x_4}{ds^2} = -x_4,$$

i.e. $d^2\mathbf{r}/ds^2 = -\mathbf{r}$, which completes the proof.

Scaling, we can get the unit vector fields defined at all points of space except the origin. Let us find symmetric polynomials S_k of the principal curvatures of the second kind of the streamlines of one of those fields, say $\mathbf{n} = \mathbf{a}/|\mathbf{a}|$. Set $\rho = \sqrt{x_1^2 + \cdots + x_n^2}$. The following theorem holds:

Theorem *Odd symmetric polynomials of principal curvatures of the second kind are zero, while even symmetric polynomials are* $S_k = \frac{1}{\rho^k} C_{(n-2)/2}^{k/2}$.

Let us show that all the principal curvatures of the second kind are purely imaginary except one of them which is zero. Denote by ξ_i the components of $\mathbf{n}$. The matrix $\left\|\frac{\partial \xi_i}{\partial x_j}\right\|$ has the form

$$\Lambda = \left\| \begin{matrix} \frac{x_2 x_1}{\rho^3} & \frac{1}{\rho} - \frac{x_1 x_1}{\rho^3} & \dots & \dots \\ \frac{x_2 x_2}{\rho^3} - \frac{1}{\rho} & -\frac{x_1 x_2}{\rho^3} & \dots & \dots \\ \dots & \dots & \dots & \dots \\ \frac{x_2 x_{n-1}}{\rho^3} & \dots & \frac{x_n x_{n-1}}{\rho^3} & \frac{1}{\rho} - \frac{x_{n-1}^2}{\rho^3} \\ \frac{x_2 x_n}{\rho^3} & \dots & \frac{x_n x_n}{\rho^3} - \frac{1}{\rho} & -\frac{x_n x_{n-1}}{\rho^3} \end{matrix} \right\|$$

To obtain a more convenient form of the characteristic equation $|\Lambda - \lambda E| = 0$ we introduce the following vectors:

$$\mathbf{l} = \frac{1}{\rho^3}\begin{pmatrix} x_1 \\ x_2 \\ \vdots \\ x_n \end{pmatrix}, \quad \tau_1 = -\begin{pmatrix} \lambda \\ \frac{1}{\rho} \\ 0 \\ \vdots \\ 0 \end{pmatrix}, \quad \tau_2 = \begin{pmatrix} \frac{1}{\rho} \\ -\lambda \\ 0 \\ \vdots \\ 0 \end{pmatrix}, \quad \dots$$

$$\dots, \quad \tau_{n-1} = -\begin{pmatrix} 0 \\ \vdots \\ 0 \\ \lambda \\ \frac{1}{\rho} \end{pmatrix}, \quad \tau_n = \begin{pmatrix} 0 \\ \vdots \\ 0 \\ \frac{1}{\rho} \\ -\lambda \end{pmatrix}.$$

We shall denote the determinant by brackets. Then the equation $|\Lambda - \lambda E| = 0$ can be expressed as

$$[x_2\mathbf{l} + \tau_1, -x_1\mathbf{l} + \tau_2, \dots, x_n\mathbf{l} + \tau_{n-1}, -x_{n-1}\mathbf{l} + \tau_n] = 0.$$

Using the rule of determinant decomposition, we obtain

$$[\tau_1, \tau_2, \dots, \tau_n] + x_2[\mathbf{l}, \tau_2, \dots, \tau_n] - x_1[\tau_1, \mathbf{l}, \tau_3, \dots, \tau_n] + \dots = 0,$$

where dots stand for the sum of determinants each of which contains only one column $\mathbf{l}$ taking the position of the column τ_i. Those determinants are similar to the two which were presented. The matrix $[\tau_1, \dots, \tau_n]$ is formed with second order matrix boxes. Therefore, its determinant is easy to find $[\tau_1, \dots, \tau_n] = \left(\lambda^2 + \frac{1}{\rho^2}\right)^{1/2}$. Next we find:

$$[\mathbf{l}, \tau_2, \dots, \tau_n] = -\left(\lambda^2 + \frac{1}{\rho^2}\right)^{(n-2)/2}\left(x_1\lambda + \frac{x_2}{\rho}\right)\rho^{-3},$$

$$[\tau_1, \mathbf{l}, \tau_3, \dots, \tau_n] = \left(\lambda^2 + \frac{1}{\rho^2}\right)^{(n-2)/2}\left(-x_2\lambda + \frac{x_1}{\rho}\right)\rho^{-3}.$$

The characteristic equation with respect to λ has the form

$$\left(\lambda^2 + \frac{1}{\rho^2}\right)^{(n-2)/2} \lambda^2 = 0.$$

The first root $\lambda = 0$ arises from the condition that the field **n** is a unit vector field. The other roots, namely $\lambda = 0$ and $\lambda = \pm\frac{i}{\rho}$, are the principal curvatures, where the imaginary roots are of multiplicity $\frac{n-2}{2}$. The theorem is proved.

Let us find the curvature vector **P** which has been introduced in Section 2.3 for the case of the unit vector field **n** in E^n:

$$\mathbf{P} = S_{n-1}\mathbf{n} + S_{n-2}\mathbf{k}_1 + S_{n-3}\mathbf{k}_2 + \ldots + \mathbf{k}_{n-1},$$

where $\mathbf{k}_i$ are defined inductively: $\mathbf{k}_1 = \nabla_{\mathbf{n}}\mathbf{n}$, $\mathbf{k}_{i+1} = \nabla_{\mathbf{k}_i}\mathbf{n}, \ldots$. As **n** we take $\mathbf{a}/\rho$. Since every streamline of the field **a** is a great circle on a sphere, $\mathbf{k}_1 = -\mathbf{r}/\rho^2$. The field is constant along the sphere radius, so that $\mathbf{k}_2 = -\nabla_{\mathbf{r}}\mathbf{n} = 0$, $\mathbf{k}_i = 0$, $i \geq 2$. Take into account that the odd symmetric polynomials are zero. So, $\mathbf{P} = S_{n-2}\mathbf{k}_1 = -\mathbf{r}/\rho^n$. By means of the vector **P** we can find the degree of mapping generated by the field $\mathbf{a}/\rho$ of the sphere S^{n-1} onto the unit sphere in E^n:

$$\theta = \frac{1}{\omega_{n-1}} \int_{S_{n-1}} (\mathbf{P}, \nu)\, dV = -1.$$

Consider now the question of the holonomicity of pairs of fields. We shall say that the distribution of the pair field **a**, **b** is holonomic if there are $(n-2)$-dimensional submanifolds orthogonal to both **a** and **b**. Otherwise we say that the distribution is non-holonomic.

In papers we also meet another definition: the distribution of planes spanned on **a** and **b** is called integrable if there are two-dimensional submanifolds tangent to those planes. The context will show in which sense this notion is used.

If we take a pair of fields **a**, **b** in E^4 then their distribution is holonomic. We can prove this fact both analytically in terms of exterior forms and geometrically. The fields **c** and **r** are orthogonal to **a** and **b**. Therefore, the planes of great circles in the sphere S^3 which are tangent to the field **c** are orthogonal to **a** and **b**. This means that the distribution of the fields **a** and **b** is holonomic in E^4.

On the other hand, the distribution of fields **c** and **r** is not holonomic in E^4. Indeed, write the corresponding Pfaff system:

$$\omega_1 = x_1\, dx_1 + x_2\, dx_2 + x_3\, dx_3 + x_4\, dx_4 = 0,$$
$$\omega_2 = -x_4\, dx_1 - x_3\, dx_2 + x_2\, dx_3 + x_1\, dx_4 = 0.$$

The exterior differentials are

$$d\omega_1 = 0, \quad d\omega_2 = 2(dx_1 \wedge dx_4 + dx_2 \wedge dx_3).$$

Set $\Delta = x_1x_3 - x_2x_4$. Exclude the differentials dx_1 and dx_2 from the Pfaff system:

$$dx_1 = -\{(x_2^2 + x_3^2)\, dx_3 + (x_3x_4 + x_1x_2)\, dx_4\}/\Delta,$$

$$dx_2 = \{(x_1x_2 + x_3x_4)\,dx_3 + (x_1^2 + x_4^2)\,dx_4\}/\Delta.$$

Substitute them into $d\omega_2$:

$$d\omega_2 = -\frac{2}{\Delta}(x_1^2 + x_2^2 + x_3^2 + x_4^2)\,dx_3 \wedge dx_4.$$

Hence, $d\omega_2$ is not zero identically by virtue of the system. Therefore, the distribution is non-holonomic. Thus, we have an example of two distributions of two-dimensional mutually orthogonal planes in E^4 such that one of them is holonomic, while the other is not.

The case considered above is exceptional. In a space of larger dimension the distribution of vector pairs is non-holonomic. Consider, for instance, the distribution of the vector pair $\mathbf{a}, \mathbf{b}$ in E^8. The corresponding Pfaff system is

$$\begin{aligned}
\omega_1 = (\mathbf{a}, d\mathbf{r}) = {} & -x_2\,dx_1 + x_1\,dx_2 - x_4\,dx_3 + x_3\,dx_4 \\
& - x_6\,dx_5 + x_5\,dx_6 - x_8\,dx_7 + x_7\,dx_8 = 0, \\
\omega_2 = (\mathbf{b}, d\mathbf{r}) = {} & -x_3\,dx_1 + x_4\,dx_2 + x_1\,dx_3 - x_2\,dx_4 \\
& + x_7\,dx_5 - x_8\,dx_6 - x_5\,dx_7 + x_6\,dx_8 = 0.
\end{aligned}$$

Let us find, for instance, the exterior form $d\omega_1$:

$$d\omega_1 = 2\{dx_1 \wedge dx_2 + dx_3 \wedge dx_4 + dx_5 \wedge dx_6 + dx_7 \wedge dx_8\}.$$

From the system $\omega_1 = 0$, $\omega_2 = 0$, we find

$$dx_1 = \sum_{\alpha=3}^{8} a_\alpha\,dx_\alpha, \qquad dx_2 = \sum_{\beta=3}^{8} b_\beta\,dx_\beta,$$

where a_α, b_β are some coefficients. After the substitution of $dx_1 \wedge dx_2$ into $d\omega_1$, the latter will contain the term $(a_3b_5 - b_3a_5)dx_3 \wedge dx_5$. In collecting similar terms this term does not vanish. Therefore, $d\omega_1 \neq 0$ by virtue of $\omega_1 = 0$, $\omega_2 = 0$, i.e. the distribution is non-holonomic.

Let us consider a more complicated system of three Pfaff equations in E^8:

$$\omega_1 = (\mathbf{a}, d\mathbf{r}) = 0, \quad \omega_2 = (\mathbf{b}, d\mathbf{r}) = 0, \quad \omega_3 = (\mathbf{c}, d\mathbf{r}) = 0.$$

We already found the exterior differential of $d\omega_1$ above. Exclude dx_1, dx_2, dx_3 from the Pfaff system and substitute into $d\omega_1$. In the Pfaff system $\omega_i = 0$, $i = 1, 2, 3$ $\mathbf{a}_j$ denotes the three-element column formed with the coefficients of dx_j. The determinant of the matrix of the system with respect to dx_1, dx_2, dx_3 has the following form:

$$\Delta = [a_1a_2a_3] = \begin{vmatrix} -x_2 & x_1 & -x_4 \\ -x_3 & x_4 & x_1 \\ -x_4 & -x_3 & x_2 \end{vmatrix} = -x_4(x_1^2 + x_2^2 + x_3^2 + x_4^2).$$

Find the coefficient of $dx_4 \wedge dx_5$ in $d\omega_1$. To do this we only need the terms with dx_4 and dx_5 in the expressions for dx_1, dx_2, dx_3. Write

$$dx_1 = \frac{1}{\Delta}(A\,dx_4 + B\,dx_5) + \cdots,$$

$$dx_2 = \frac{1}{\Delta}(C\,dx_4 + D\,dx_5) + \cdots,$$

$$dx_3 = \frac{1}{\Delta}(E\,dx_4 + F\,dx_5) + \cdots,$$

where $A, \ldots, F$ have the form

$$\begin{aligned} A &= -[a_4a_2a_3], & B &= -[a_5a_2a_3], \\ C &= -[a_1a_4a_3], & D &= -[a_1a_5a_3], \\ E &= -[a_1a_2a_4], & F &= -[a_1a_2a_5]. \end{aligned}$$

Substituting dx_1, dx_2, dx_3 into $d\omega_1$, we obtain

$$d\omega_1 = \frac{1}{\Delta^2}(AD - BC - F\Delta)dx_4 \wedge dx_5 + \cdots,$$

where the dots stand for terms without $dx_4 \wedge dx_5$. Set $U = (x_1^2 + x_2^2 + x_3^2 + x_4^2)$. Then

$$A = -\begin{vmatrix} x_3 & x_1 & -x_4 \\ -x_2 & x_4 & x_1 \\ x_1 & -x_3 & x_2 \end{vmatrix} = -x_1U,$$

$$C = -\begin{vmatrix} -x_2 & x_3 & -x_4 \\ -x_3 & -x_2 & x_1 \\ -x_4 & x_1 & x_2 \end{vmatrix} = -x_2U,$$

$$B = -\begin{vmatrix} -x_6 & x_1 & -x_4 \\ x_7 & x_4 & x_1 \\ -x_8 & -x_3 & x_2 \end{vmatrix} = x_2x_4x_6 + x_1^2x_8 - x_3x_4x_7 + x_4^2x_8 + x_1x_3x_6 + x_1x_2x_7,$$

$$D = -\begin{vmatrix} -x_2 & -x_6 & -x_4 \\ -x_3 & x_7 & x_1 \\ -x_4 & -x_8 & x_2 \end{vmatrix} = x_2^2x_7 - x_1x_4x_6 + x_3x_4x_8 + x_4^2x_7 + x_1x_2x_8 + x_2x_3x_6,$$

$$F = -\begin{vmatrix} -x_2 & x_1 & -x_6 \\ -x_3 & x_4 & x_7 \\ -x_4 & -x_3 & -x_8 \end{vmatrix} = -x_2x_4x_8 + x_1x_4x_7 + x_3^2x_6 + x_4^2x_6 + x_1x_3x_8 + x_2x_3x_7.$$

After the substitution, we obtain

$$AD - BC - F\Delta = x_4x_6U^2.$$

Therefore at all points where x_4, x_6 and U are different from zero, the coefficient of $dx_4 \wedge dx_5$ in $d\omega_1$ is not equal to zero. If we set $dx_6 = dx_7 = \delta x_6 = \delta x_7 = 0$ then all the forms $dx_j \wedge dx_6$ and $dx_j \wedge dx_7$ are zero. So, $d\omega_1 \not\equiv 0$ by virtue of the system, i.e. the Pfaff system is not integrable.

Each of the vector fields constructed is a Killing field on a sphere. We get the proof from the definition. Such a field ξ defines an infinitesimal mapping $\mathbf{r} \to \bar{\mathbf{r}}$ of a Riemannian manifold onto itself under which $d\bar{\mathbf{r}}^2$ differs from $d\mathbf{r}^2$ by the infinitesimal of a higher order than $(\delta t)^2$. Put the point of the position vector $\mathbf{r}$ into correspondence with the point of the position vector $\bar{\mathbf{r}} = \frac{\mathbf{r}+\xi\delta t}{\sqrt{1+(\delta t)^2}}$ in a unit sphere. As the metric of the sphere is an induced one from the ambient Euclidean space, we can take the differential of the square of $\bar{\mathbf{r}}$ in Euclidean space to find this metric:

$$d\bar{\mathbf{r}}^2 = \{d\mathbf{r}^2 + 2(d\mathbf{r}, d\xi)\,\delta t + d\xi^2(\delta t)^2\}/(1 + (\delta t)^2),$$

where $(d\mathbf{r}, d\xi)$ means the scalar product in Euclidean space. Take one of the fields constructed as the field ξ. It is sufficient to check that $(d\mathbf{r}, d\xi) = 0$. Each field ξ at a point with the position vector $\mathbf{r}$ is obtained from $\mathbf{r}$ by component transpositions and sign changes. More precisely, if x_i takes the j-th place then $-x_j$ takes the i–th place. Therefore, the products $dx_i\,dx_j$ meet twice in the expression of $(d\mathbf{r}, d\xi)$ — once with a negative sign and once with a positive sign. Therefore, $(d\mathbf{r}, d\xi) = 0$ and, as a consequence, ξ is a Killing vector field.

2.20 On the Family of Surfaces Which Fills a Ball

Let D^n be a ball in Euclidean space E^n covered with a one–parametric family of surfaces (i.e. through each point of the ball one and only one surface passes). We shall suppose that each surface of the family corresponds to a unique value of the parameter t from $[a, b]$ and vice versa, and also that this correspondence is continuous. Moreover, suppose that each surface is connected and divides the ball into two bodies such that in varying t from a to b the volume of one of those bodies varies from zero to the volume of the whole ball D^n. We shall say that such a *family fills the ball regularly*. Nevertheless, the surfaces themselves may have singularities. For instance, we may take as such a family the set of circular cones with a common axis and the same angle about the vertex if this vertex is within the ball and completed with the family of frustums of the cones if the vertex is out of the ball. Two examples are: the family of coaxial cylinders truncated by the ball; the family of level surfaces of some differentiable function $\varphi(x_1, \ldots, x_n)$ with a non-zero gradient within the ball. The surfaces which correspond to extremal values of t can be degenerated into a point or a curve.

We can pose the following simple question: *is there a surface among those surfaces having an $(n-1)$-dimensional area greater than or equal to the area the diametral section of D^n?*

For $n = 2$ this question asks the lengths of curves which cover a disk of radius R. Among the curves of this family there is the curve which passes through the center of the disk. Since this curve comes to the boundary circle, the length of this curve $\geq 2R$. For the n-dimensional case we prove the following assertion.

Theorem *Let us be given a family of surfaces which fills a ball D^n of radius R regularly. Then there is a surface of the family with an area greater than or equal to*

$R^{n-1}C(n)$, *where* $C(n) = \omega_{n-1}(2^{1/n} - 1)/2$ *is constant and* ω_{n-1} *is the area of a unit* $(n-1)$*-dimensional sphere.*

Let V_1 and V_2 be the volumes of the bodies D_1 and D_2 which are the parts into which the surface divides the ball D^n. The boundary of each D_i consists of the domain in a boundary sphere having an $(n-1)$-dimensional area F_i and the surface under consideration has area F. Therefore, the boundary area of D_i is $F + F_i$. Consider the isoperimetric inequality with respect to each D_i:

$$F_i + F \geq n\sigma_n^{1/n} V_i^{(n-1)/n}, \quad i = 1, 2,$$

where σ_n is the volume of a unit ball in E^n. Adding these inequalities, we get

$$2F \geq n\sigma_n^{1/n}\left(V_1^{(n-1)/n} + V_2^{(n-1)/n}\right) - (F_1 + F_2).$$

Since the family of surfaces fills the ball regularly, the volume of D_1 varies continuously from zero to the volume of D^n when the parameter t varies from a to b. Therefore, there is a surface which splits the ball into two bodies of equal volume, i.e. in this case $V_1 = V_2 = R^n\sigma_n/2$. Also, note that $F_1 + F_2 = \omega_{n-1}R^{n-1}$ is the area of the whole boundary sphere. Hence,

$$2F \geq 2^n\sigma_n R^{n-1}2^{(n-1)/n} - \omega_{n-1}R^{n-1}.$$

Note that $n\sigma_{n-1} = \omega_{n-1}$. Therefore, $2F \geq \omega_{n-1}R^{n-1}(2^{1/n} - 1)$. The theorem is proved.

Consider now the impact of the curvature of family surfaces on the size of the domain of definition. Suppose that the field $\mathbf{n}$ of normals of surfaces is given in D^n and is differentiable. Suppose that the second symmetric polynomial S_2 of each surface is greater than some fixed positive number K_0. Since $S_1^2 \geq \frac{2(n-1)}{n-2}S_2$, the divergence of a normal field is bounded from below as $\operatorname{div}\mathbf{n} \geq \sqrt{\frac{2(n-1)}{n-2}K_0}$ $\left(\text{or from above as } \operatorname{div}\mathbf{n} \leq -\sqrt{\frac{2(n-1)}{n-2}K_0}\right)$. If we integrate this inequality over the ball D^n, we obtain

$$\omega_{n-1}R^{n-1} \geq \sigma_n R^n\sqrt{\frac{2(n-1)}{n-2}K_0}.$$

Hence, *the radius of a ball, where a regular family of surfaces with a symmetric function* $S_2 \geq K_0 > 0$ *can be given, is bounded from above as*

$$R \leq \frac{n}{\sqrt{K_0}}\sqrt{\frac{(n-2)}{2(n-1)}}.$$

This assertion can be extended to the case of Riemannian space, but note that in Riemannian space it is impossible, in general, to get the upper bound on the radius of a ball for an arbitrary K_0 where the regular family of surfaces with a second symmetric function $S_2 \geq K_0 > 0$ exists. Indeed, in Lobachevski space L^3 with curvature

-1 every geodesic sphere has the normal curvatures ≥ 1. Hence, we can take a ball, however large the radius, filled with a regular family of surfaces, namely with pieces of geodesically parallel spheres of $S_2 \geq 1$. But if $K_0 > 1 + \varepsilon$, $\varepsilon > 0$ then it is possible to find the upper bound on the radius of a ball $D^3 \subset L^3$. However, we fix the ball radius to be 1 and watch the influence of K_0 on the curvature of the space. Integrating the inequality $\operatorname{div} \mathbf{n} \geq C\sqrt{K_0}$ over D^3 in Riemannian space, we obtain the estimate $\sqrt{K_0} \leq CS/V$, where S is the area of boundary sphere, V the volume of D^3, $C = \text{const}$. The right-hand side of this inequality can be estimated from above in terms of the maximum Q of sectional curvatures of Riemannian space (see [54]). Therefore, $\sqrt{K_0} \leq f(Q)$, where f is some monotone increasing function of Q. If we interpret the curvature of the family of surfaces as some loading on the domain of space then we may say that if this loading is greater than some critical value then the metric of ambient space will change, i.e. the Euclidean space will change into the curved one; the value of curvature will be determined to some extent by the value of loading. This influence is similar to the deformation of a flat membrane under the action of interior tension. The membrane loses its flatness, swells into space and gains curvature. The estimate obtained means that if K_0 increases, then after some critical point, the space curvature Q also increases.

Consideration of the families of the surface in a ball leads to some problems. Let us formulate two of them:

(1) *Suppose that in a ball $D^3 \subset E^3$ of radius R there is a regular family of C^2-regular surfaces having negative Gaussian curvature $K \leq -1$. Does the estimate from above on R hold?*
Observe that in Section 1.14 the existence of such an estimate has been proved, provided that the curvature of the normal vector field streamlines is bounded from above.

(2) *Suppose that in a ball D^n the regular vector field* $\mathbf{n}$ *is given. Is there a regular homotopy which preserves the vector field on the boundary of the ball and turns it into the integrable one?*

A more general formulation involves a domain G in Riemannian space and a given distribution of planes.

The latter problem is connected with one posed by Reeb, Chern and others: suppose that on a differential manifold of dimension n the field of k-dimensional ($k \leq n$) tangent planes is given. Under which conditions is this field homotopic to the totally integrable one? Wood [99] proved that every field of 2-planes which is transversally orientable is homotopic to the totally integrable one. The first example of a field which is not homotopic to the totally integrable one was given by Bott [90]. He found some necessary conditions for the field of planes to be homotopic to the totally integrable one.

Note the following result obtained by Thurston [100]: every $(n-1)$-plane field on a closed manifold M^n is homotopic to the tangent plane field of a C^∞, codimension-one foliation.

2.21 Summary

We have studied the local and global geometrical characteristics of vector fields in n-dimensional Riemannian space. We have also found some relations between these characteristics and topology, the theory of ordinary differential equations and fluid mechanics. In general, the vector field was assumed to be regular and to belong to classical class C^2. However, in certain cases the field under consideration could have singular points, which were interpreted as sources of curvature of definite power. In recent decades, the ideas of non-holonomic geometry combined with the theory of generalized functions has allowed a theory of currents on manifolds and varifolds to be constructed. Nevertheless, further development of the classical theory of vector fields is also very important. I hope that this book will help readers to prepare themselves for new discoveries in this area of mathematics.

References

[1] Voss, A. Geometrische Interpretation Differetialgleichung $P\,dx + Q\,dy + R\,dz = 0$. *Math. Ann.* **16** (1880) 556–559.
[2] Voss, A. Zur Theorie der algemeinen Punkt-ebensysteme. *Math. Ann.* **23** (1884) 43–81.
[3] Voss, A. Zur Theorie der algebraischen Differetialgleichungen erster Ordung erster Grades. *Math. Ann.* **23** (1884) 157–180.
[4] Voss, A. Theorie der rationalen algebraischen Punkt-Ebenen-Systeme. *Math. Ann.* **23** (1884) 359–411.
[5] Voss, A. Über die Differentialgleichungen der Mechanik. *Math. Ann.* **25** (1885) 258–286.
[6] Lilienthal, R. Über kurzsten Integralkurven einer Pfaffschen Gleichungen. *Math. Ann.* **52** (1899) 417–432.
[7] Lie, S. *Geometrie der Berührungstransformationen.* Bd. I, Leipzig: B. G. Teubner, 1896.
[8] Rogers, G. Some Differential properties of the orthogonal trajectories of a congruence of curves. *Proc. Irish Academy* **29** A, No. 6 (1912).
[9] Dauthevill, S. Sur les systemes non holonomes. *Bull. Soc. Math. France* **37** (1909).
[10] Darboux, G. Sur differentes propriétés des trajectories orthogonales d'une congruence des courbes. *Bull. Soc. Math. France* **36** (1912) 217–232.
[11] Darboux, G. *Lecons sur les systems orthogonaux et les coordones curvilignes.* Paris: Gauthier-Villars, 1910.
[12] Carathéodory, C. Untersuchungen úber die Grundlagen der Thermodynamik. *Math. Ann.* **67** (1909) 355–386.
[13] Bianchi, L. *Vorlesungen úber Differentialgeometrie.* Leipzig, Berlin: Druck und Verlag von B. G. Teubner, 1910.
[14] Chaplygin S. A. *Investigation on the dynamics of non-holonomic systems.* Moscow: GITTL, 1949. [Russian]

[15] Egorov, D. F. *Proceedings in differential geometry*. Moscow: Nauka, 1970. [Russian]

[16] Sintsov D. M. *Proceedings in non-holonomic geometry*. Kiev: Vistshaya Shkola, 1972. [Russian]

[17] Blank, Ya. P. Über eine geometrische Deutung der Integrabilitäts bedingung der Pfaffschen Differntialgleichung. *Notes of Kharkov Math. Soc.* **13** (1936) 75–81.

[18] Schouten, J. A. Über nicht holonome Üebertragungen in einer L^n. *Math. Zeit.* **30** (1929) 149–172.

[19] Schouten, J. A. and von Kampen E.R. Zur Einbettungs und Krummungstheorie nicht holonomie Gebilde. *Math. Ann.* **103** (1930) 752–783.

[20] Vrănceanu, G. Sur les espases non holonomes. *Comp. Rend.* **183** (1926) 835–854.

[21] Vrănceanu, G. Etude des espaces non holonomes. *J. Math.* **13** (1934) 131–132.

[22] Vrănceanu G. La theorie unitaire des champs et les hypersurfaces non holonomes. *Comp. Rend.* **200** (1935) 2056.

[23] Vrănceanu, G. Sur les systemes integres d'un systeme de Pfaff. *Bull. Math.* **35** (1934) 262.

[24] Vrănceanu, G. *Lectii de geometrie differentială*. Bucharest, **II** 1951.

[25] Krein, M. Über den Satz von "Curvature integra". Izv. Kazan. *Phys.-Math. Soc.* **3** (1928) Ser. 3.

[26] Lopshits, A. M. The nonholonomic relations system in many-dimensional Euclidean space. *Proc. Semin. Vector Tensor analysis*, **4** (1937) 302–332. [Russian]

[27] Goursat, E. *Lecons sur le problemm de Pfaff.* Paris: Hermann (1922).

[28] Griffin, M. Invariants of Pfaffian systems. *Trans. Ann. Math. Soc.* **35** (1935) 936.

[29] Rashevski, P. K. On the connectability of any two points of a totally nonholonomic space with admissible curve. *Sci. notes of Ped. Inst., Ser. phys.-math.* **2** (1938) 83–94. [Russian]

[30] Vagner, V. V. On the geometrical interpretation of the curvature vector of a non-holonomic V_3^2 in three-dimensional Euclidean space. *Math. Sb.* **4** (1938) 339–356.

[31] Vagner, V. V. Differential geometry of non-holonomic manifolds. *Proc. Semin. Vector Tensor analysis*, **II–III** (1925) 269–314. [Russian]

[32] Vagner, V. V. Differential geometry of non-holonomic manifolds. *VIII Int. competition on Lobachevski prize, Kazan* (1940) 195–262. [Russian]

[33] Vagner, V. V. Geometry of $(n-1)$-dimensional nonholonomic manifold in n-dimensional space. *Proc. Semin. Vector Tensor analysis*, **5** (1941) 173–225. [Russian]

[34] Vagner, V. V. Geometrical interpretation of moving nonholonomic dynamical systems. *Proc. Semin. vector tensor analysis*, **5** (1941) 301–327. [Russian]

[35] Vagner, V. V. The theory on circle congruences and the geometry of nonholonomic V_3^2. *Proc. Semin. Vector Tensor analysis*, **5** (1941) 271–283. [Russian]

[36] Bushgens, S. Geometry of vector fields. *Rep. Acad. Sci. USSR, Ser. Mat.* **10** (1946) 73–96. [Russian]

[37] Bushgens, S. Critical surface of Adiabatic flow. *Sov. Math. Dokl.* **LVIII** (1947) 365–368. [Russian]

[38] Bushgens, S. Geometry of stationary flow of an ideal incompressible liquid. *Rep. Acad. Sci. USSR, Ser. Mat.* **12** (1948) 481–512. [Russian]

[39] Bushgens, S. On a parallel displacement of vector fields *Vestnik MSU* No. 2 (1950) 3–6. [Russian]

[40] Bushgens, S. On ideal incompressible liquid streamlines. *Sov. Mat. Dokl.* **78** (1951) 837. [Russian]

[41] Bushgens, S. Geometry of adiabatic flow. *Sci. notes NSU, Ser. Mat.* **148** (1951) 50–52. [Russian]

[42] Bushgens, S. On the streamlines II. *Sci. Math. Dokl.* **LXXXIV** (1952) 861. [Russian]

[43] Pan, T.K. Normal curvature of a vector field. *Amer. J. Math.* **74** (1952) 955–966.

[44] Coburn, N. Intrisic relations satisfied by vorticity and velocity vectors in fluid theory. *Michigan Math. J.* **1** (1954) 113–130.

[45] Coburn, N. Note on my paper "Intrisic relations...". *Michigan Math. J.* **2** (1954) 41–49.

[46] Cheorghiev, Ch. *Studii si cercetări. Acad. R.P.Romania, Stiin matem. Filiala Jasi Math.* Sec. 1 **8** No. 2 (1957).

[47] Sluhaev, V. Non-holonomic manifolds V_n^m of zero external curvature. *Ukr. Geom. Sb.* **4** (1967) 78–84. [Russian]

[48] Sluhaev, V. Towards the geometrical theory of liquid stationary motion. *Sov. Mat. Dokl.* **196** No. 3 (1971) 549–552. [Russian]

[49] Sluhaev, V. Geometrical theory of ideal liquid stationary motion. *Novosibirsk, Numer. methods of continuous medium mechanics.* **4** No. 1 (1973) 131–145. [Russian]

[50] Sluhaev, V. *Geometry of vector fields.* Tomsk: Tomsk University Press, 1982. [Russian]

[51] Neimark, Yu. and Fufaev, N. *Dynamics of non-holonomic systems.* Moscow: Nauka, 1967. [Russian]

[52] Krasnosel'ski, M. *Topological methods for nonlinear integral equations theory.* Moscow: Gostehizdat, 1956. [Russian]

[53] Krasnosel'ski, M., Perov, A., Zabrejko, P. and Povolotski, A. *Vector fields on a plane.* Moscow: Fizmatgiz, 1963. [Russian]

[54] Aminov, Yu. Vector field curvature sources. *Math. Sb.* **80** (1969) 210–224. [Russian]

[55] Aminov, Yu. The divergent properties of vector field curvatures and the families of surfaces. *Math. Zametki.* **3** (1968) 103–111. [Russian]

[56] Aminov, Yu. Some global aspects in the geometry of vector fields. *Ukr. Geom. Sb.* **8** (1970) 3–15. [Russian]

[57] Aminov, Yu. On the behavior of vector field streamlines near the cycle. *Ukr. Geom. Sb.* **12** (1972) 3–12. [Russian]

[58] Aminov, Yu. On an energetic condition for the existence of rotation. *Math. Sb.* **86** (1971) 325–334. [Russian]
[59] Aminov, Yu. On estimates of the diameter and volume of a submanifold of Euclidean space. *Ukr. Geom. Sb.* **18** (1975) 3–15. [Russian]
[60] Aminov, Yu. The holonomicity condition for the principal curvatures of a submanifold. *Math. Zametki.* **41** (1987) 543–548. [Russian]
[61] Aminov, Yu. The multidimensional generalization of the Gauss–Bonnet formula to vector fields in Euclidean space. *Math. Sb.* **134** (1987) 135–140. [Russian]
[62] Efimov, N. The investigation of the unique projection of a negatively curved surface. *Sov. Mat. Dokl.* **93** (1953) 609–611. [Russian]
[63] Efimov, N. Non-existence of a complete regular surface of negative upper bound for Gaussian curvature in 3-dimensional Euclidean space. *Sov. Mat. Dokl.* **150** (1963) 1206–1209. [Russian]
[64] Efimov, N. The occurrence of singular points on negatively curved surfaces. *Math. Sb.* **64** (1964) 286–320. [Russian]
[65] Efimov, N. Differential criteria for some mappings to be homeomorphisms with application to the theory of surfaces. *Math. Sb.* **76** (1968) 489–512. [Russian]
[66] Whitehead, J. H. C. An expression of Hopf's invariant as an integral. *Proc. Nat. Acad. Sci. USA* **33** (1947) 117–123.
[67] Haefliger, A. Sur les feuilletages analytiques. *Comp. Rend.* **242** (1956) 2908–2910.
[68] Godbillon, C. and Vey J. Un invariant des feulletages des codimension 1. *Comp. Rend.* **273** (1971) 92–95.
[69] Reinhart, B. L. and Wood, J. W. A metric formula for the Godbillon–Vey invariant for foliations. *Proc. Amer. Math. Soc.* **38** (1973) 427–430.
[70] Tamura, I. Every odd dimensional sphere has a foliation of codimension one. *Comm. math. Helv.* **47** (1972) 164–170.
[71] Tamura, I. *Topology of foliations.* Moscow: Mir, 1979. [Russian]
[72] Pontryagin, L. *Smooth manifolds and their applications to homotopy theory.* Moscow: Nauka, 1976. [Russian]
[73] Novikov, S. Topology of foliations. *Proc. Moscow Math. Soc.* **14** (1965) 248–278. [Russian]
[74] Novikov, S. The generalized analytic Hopf invariant. Multivalued functionals. *Uspehi Mat. Nauk*, **36** (1984) 97–106. [Russian]
[75] Gantmacher, F. *Matrix Theory.* Moscow: Nauka, 1988. [Russian]
[76] Kosevich, A., Ivanov, B. and Kovalev A. Nonlinear waves in magnetism. Dinamical and topological solitons. Kiev: Naukova dumka, 1983. [Russian]
[77] Whitney, H. *Geometrical theory of integration.* Moscow: IL, 1960. [Russian]
[78] Hu, S. *Homotopy Theory.* Moscow: Mir, 1964. [Russian]
[79] Rogovoj, M. Towards the metric theory of non-holonomic hypersurfaces of n-dimensional space. *Ukr. Geom. Sb.* **5–6** (1968) 126–138. [Russian]
[80] Rogovoj, M. Towards the differential geometry of a non-holonomic hypersurface. *Ukr. Geom. Sb.* **7** (1970) 98–108. [Russian]
[81] Rogovoj, M. On the osculating hypersurfaces of a non-holonomic manifold V_n^{n-1} in P_n. *Ukr. Geom. Sb.* **18** (1975) 116–125. [Russian]

[82] Rogovoj, M. On the curvature and torsion of a non-holonomic hypersurface. *Ukr. Geom. Sb.* **7** (1970) 109–113. [Russian]
[83] Rogovoj, M. Towards the projective-differential geometry of non-holonomic surface. *Ukr. Geom. Sb.* **8** (1970) 112–119. [Russian]
[84] Rogovoj, M. Projective characteristics of the non-holonomic manifold V_n^{n-1}. *Ukr. Geom. Sb.* **11** (1971) 83–86. [Russian]
[85] Lumiste, U. Towards the theory of manifold of planes in Euclidean space. *Sci. Notes Tartu Univ.* **192** No. 6 (1966) 12–46. [Russian]
[86] Lumiste, U. and Gejdel'man, R. Geometry of family of m-dimensional spaces in n-dimensional spaces. *Proc. fourth All-Union Congr. of Math., Leningrad,* **2** (1964) 201–206. [Russian]
[87] Laptev, G. and Ostianu, N. The distribution of m-dimensional elements in spaces of projective connection. *Proc. Geom. Semin., Viniti* **3** (1971) 49–94. [Russian]
[88] Ostianu, N. The distribution of hyperplane elements in projective space. *Proc. Geom. Semin., Viniti,* **4** (1973) 71–120. [Russian]
[89] Bott, R. Vectorfields and characteristic numbers. *Michigan Math. J.* **14** (1967) 231–244.
[90] Bott, R. A residue formula for holomorphic vector fields. *J. Dif. Geom.* **1** (1967) 231–244.
[91] Baum, B. F., Cheeger J. Infinitesimal isometries and Pontryagin numbers. *Topology* **8** (1969) 173–193.
[92] Kantor, B. Towards the problem of the normal image of a complete surface of negative curvature. *Math. Sb.* **82** (1970) 220–223. [Russian]
[93] Eckmann, B. Gruppentheoretischer Beweis des Satzes von Hurwitz–Radon. *Comm. Math. Helv.* **15** (1942/43) 358–366.
[94] Adams, J. F. Vector fields on spheres. *Ann. Math.* **75** (1962) 603–632.
[95] Stepanov, V. *A course in differential equations.* Moscow, Gostehizdat, sixth edition 1953. [Russian]
[96] Rumjantsev, V. On a integral principal for non-holonomic systems. *PMM* **46** (1982) 3–12. [Russian]
[97] Rumjantsev, V. and Karapetjan, A. *Stability in a moving non-holonomic system. Itogi Nauki i Tehn., General Mech. M. VINITI* **3** (1976) 5–42. [Russian]
[98] Raby, G. Invariance des classes de Godbillon–Vey C^1-diffeomorphismes. *Ann. Inst. Fourier.* **38** (1988) 205–213.
[99] Wood, J. Foliations on 3-manifolds. *Ann. Math.* **89** (1969) 336–358.
[100] Thurston, W.P. Existence of codimension-one foliations. *Ann. Math.* **104** (1976) 249–268.
[101] Fuks, D. Foliations. *Itogi Nauki, Seriya: Algebra,Topologiya, Geometriya.* **18** (1981) 151–213 [Russian]. Translation: *J. Soviet Math.* **18** (1982) 255–291.
[102] Lawson, H.B. Foliations. *Bull. A.M.S.(3)* **80** (1974) 369–418.
[103] Ferus, D. Totally geodesic foliations. *Math. Ann.* **188** (1970) 313–316.
[104] Rovenskii, V. Totally geodesic foliations. *Sibir. Math. J.* **23** (1982) 217–219.
[105] Tanno, S. Totally geodesic foliations with compact leaves. *Hokkaido Math. J.* **1** (1972) 7–11.

[106] Abe, K. Application of a Riccatti type differetial equation to Riemannian manifold with totally geodesic distribution. *Tohoku Math. J.* **25** (1973) 425–444.

[107] Johnson, D. and Whitt, L. Totally geodesic foliations. *J. Diff. Geom.* **15** (1980) 225–235.

[108] Johnson, D. and Naveira, A. A topological obstruction of the geodesibility of a foliation of odd dimension. *Geom. Dedicata.* **11** (1981) 347–357.

[109] Gromov, M. Stable mappings of foliations into manifolds. *Izv. Acad. Nauk. SSSR. Ser. Mat.* **33** (1969). 707–734.

[110] Haefliger, A. Some remarks on foliations with minimal leaves. *J. Dif. Geom.* **15** (1980). 269–284.

[111] Hass, J. Minimal surfaces in foliated manifolds. *Comm. Math. Helv.* **61** (1986). 1–32.

[112] Solomon, B. On foliations of R^{n+1} by minimal hypersurfaces. *Comm. Math. Helv.* **61** (1986) 7–83.

[113] Cairns, G. Feuilletages totalement geodesiques de dimension 1, 2 or 3. *C. R. Acad. Sci. Paris.* **298** (1984) 341–344.

[114] Tondeur, P. *Foliations on Riemannian manifolds.* Universitext, Springer-Verlag, New York, 1988.

Subject Index

Author Index